Living - Participating - Growing Old

Anton Amann

Living - Participating - Growing Old

Assumptions and Certainties

 Springer

Anton Amann
Universitat Wien
Wien, Austria

ISBN 978-3-658-39680-0   ISBN 978-3-658-39681-7   (eBook)
https://doi.org/10.1007/978-3-658-39681-7

I had planned to do a lot of work, but nothing came of it, and I had done a lot of things I hadn't thought of; so this is really living life.
(Johann W. v. Goethe to Friedrich A. Wolf 16.12.1808)

Günter Dux
dedicated

Preface

Assumptions (hypotheses) are of a scientific nature if they are methodologically justified, certainties are of a practical nature if they stem from the experience of success. All the considerations in this book are oriented towards these two perspectives. Sociological opinion, i.e. knowledge that is pretended to be sociological but lacks that sociological quality of its own that allows social things to emerge from thought, action and speech, is irrelevant. This meaning seldom grasps the matter at hand and usually misses the mark. Moreover, it is true that the social cannot be reduced to psychological events, nor is it exhausted in anonymous structures; it is the acting people who, with the help of the existing world, constructively produce their own world in a new and changing way, even if they may sometimes miss what the consequences of their actions will be.

To stand in the middle of life, to look into the future together with others with confidence, and to participate in what is happening, that is what almost everyone wishes for. The digital "openthesaurus.de" offers the following synonyms: "to participate in life, to be in the midst of life, to shape one's life, to take life in one's hands, to know how to master life, to know how to take life, to be fit for life"; the common associations are: "(to understand) the art of living, to know how to take life, to be active, not only to sit in front of the television, to undertake something". Truly a multifaceted task. But, what enables people to participate in life, what individual and structural moments make participation shape or fail, and according to what ideas is it evaluated? These three questions hold the relevant landmarks for the journey into an area of very heterogeneous research traditions, and the various possible answers will show that participation is neither unquestionably given nor does it come about on its own. From earliest childhood, even before birth, it prepares itself, forms itself and shapes itself. It is essential to gain clarity about what life means and what participation means. Its goals, contents and ways of shaping are manifold and change in history and in the individual course of life for the most diverse reasons, whereby participation in old age means its own specification of the conditions contained in the three questions mentioned at the beginning.

These specification suggest that a discussion that wants to start without presuppositions at what is simply in front of the eyes of observation as participation would have to go astray. There are no facts that "speak for themselves"; they only become findings through interpretation within the framework of binding conceptions. At least this is true from an epistemological point of view, as I pursue it here. It rests, as will be shown later, on the thesis that the forms of social life are not based in any way on idealistic presuppositions or innate forms of behaviour that cannot be further explored but are constantly created and also changed by people themselves in a constructive way. Concepts such as biological fate, a better race, extra-human forces, anonymous powers, or what is supposedly always valid prove to be of little use or even absurd in substance for sociological analysis from this perspective. The book is an attempt to contribute essential elements to a holistic view of participation.

Participation in life, in its infinite richness of facets, is nothing other than an expression of the fullness of interaction of the forms of life intentionally constructively created by human beings in their history. In view of this fact, I am attempting to ground participation sociologically anthropologically, since the conceptualizations employed in the usual empirical surveys have so far seemed to me to be rather unsatisfactory.

Wien, Austria   Anton Amann

Contents

Part I

Idea-Historical and Epistemological Approach

Life and the Acting Human Being

What life can mean, for all its confusion and despite all the contradictions provoked by previous answers, cannot be separated from a conception: It can only be conceived as the result of evolution, and if we consider – in a narrower understanding – human life, only as the emergence of the species Sapiens from the genus Homo of the family of the apes. If the emergence of human life as an evolutionary fact is coupled with the notion of social life, i.e. the constructive bringing forth of human life forms by human beings themselves (what some call cultural evolution), then on the one hand the path is opened on which further discussions can proceed, but also the set of all traps is set up into which one can fall on this path. For what is at stake is nothing less than to ascertain how participation, on the one hand, is tied back to the phylogenesis of the species Sapiens, but, on the other hand, must be acquired and practiced anew by each human being in ontogeny, relying on environment. We will encounter well known pairings along the way such as: self-socialization theory and social theory, instinct theory and "mind" theories, individualism and collectivism, coercion and autonomy, etc. We will encounter political aporias and social illusions, and possibly we will get the chance to learn that many a dispute about alleged ultimate truths is a dispute about very ephemeral constructions of thought, whose relative durability can be decided not only by arguments about content but also by power relations. Scientific knowledge does not come about outside of social relations and unaffected by them. That egoism or the will to cooperate are inevitably innate, or that people with black skin colour possess less intelligence than those with white skin colour, are views that make themselves suspect of disorderly origins.

In order to gain an idea of how the reconstruction of the social behaviour "participation" might be thought of, I will begin with a review of some developments

A. Amann, *Living - Participating - Growing Old*, https://doi.org/10.1007/978-3-658-39681-7_1

in philosophical anthropology, attempt an overview of the thoughts of Günter Dux, which seem to me to be undoubtedly the most sociologically fruitful at present for this particular topic, and finally seek to argue why the relation of participation to evolution is justified. The analyses and reports of results that follow are of a historical-concrete and empirical-analytical nature and constitute the part that is social gerontological in the narrower sense. The totality of empirical research results is intentionally very rich in order to better distinguish between widespread opinions and scientific findings.

1.1   Reservoir of Ideas in Philosophical Anthropology[1]

The reason for going into more detail here about traditions in philosophical anthropology is, on the one hand, to track down the still highly effective notions and ideas that continue to influence thinking about participation in life and the formation of its cultural foundations, explicitly or unspokenly, to this day (cf. e.g. Blumenberg 1979; Bayertz 2012; Fagan 2012; Gronau 2016; Eilenberger 2018, p. 151). They appear in sometimes easily, sometimes hardly recognizable forms in sociological and psychological theories, also in historically oriented accounts (cf. MacGregor 2017), they are an integral part of philosophical and political science discourses, and they determine to a considerable extent the image of man behind ideological-political conceptions, for example. Another and even more important reason is to present considerations by Günter Dux, from which he has developed his "historico-genetic theory of culture" in a clear break with older views, which in turn again allows helpful use for the definition of participation as I understand it. This connection allows for a decidedly deeper reading of the idea of participation than has previously been cultivated in social research. The concept of involvement/disengagement (see the further discussion) can be directly linked to the constructive development of life forms by human beings. Max Scheler has pointed out that three traditions of thought usually play a role as sources of possible answers in this context of ideas. The Judeo-Christian tradition with creation, paradise and the fall of

---

[1] I have chosen the authors assembled in this chapter because I believe that they are among those who (along with others such as Immanuel Kant and Wilhelm v. Humboldt or Gottfried W. Leibniz and Friedrich W. Schelling as well as Friedrich Nietzsche) most clearly represent the reservoir of ideas that influenced thinking in Europe for over 200 years on the question of man's place in the world, before it underwent decisive changes from the 1970s onwards as a result of new findings in palaeontology, archaeology, brain research and evolutionary biology.

man; the Greek-ancient tradition, which establishes man as a rational being (λόγος: reason; φρόνησις: prudence); finally, the tradition of modern natural sciences and genetic psychology, which understand man as a late result of evolution (Scheler 1988, p. 9). Since especially the theory of the "deficient being" has drawn impulses from the last two traditions, a theory that still finds advocacy from time to time today, it is necessary to go back to an early point of reference: Johann G. Herder.

1.1.1   Johann G. Herder: The Great Forerunner

Twenty-eight years before Georg W. F. Hegel, who saw Christian consciousness as the cornerstone of Western culture, Johann G. Herder (1744–1803) died, who, among many other admirable things he left behind for posterity, laid the foundation for a new anthropologically based conception of man, the core of which can still claim interest today. Alongside Michel de Montaigne, Giambattista Vico and Jean-Jacques Rousseau, he is regarded as a precursor of cultural philosophy. Max Scheler's and Arnold Gehlen's conceptions of culture are directly related to Johann G. Herder, and even Helmuth Plessner's can be seen as loosely connected to him. Johann G. Herder's exposition is based on the idea of a philosophical anthropology that directs the natural and historical philosophical view to the empirical results of the individual sciences. In sometimes almost poetic language he sums up the knowledge of the time in a clear summary. "I dare, since I am not a dissector, to follow the perceptions of great dissectors in a pair of examples" (Herder 1784, p. 75),[2] he tries to integrate findings of the individual sciences of his guiding idea of a philosophy of the history of mankind.

He finds a first approach in the difference between man and animal, which was already established by Aristotle, still plays an essential role today and is only occasionally seen as worth overcoming.[3] Johann G. Herder thus observes man in

---

[2] The original spelling has been retained.

[3] In the meantime, however, the human/animal difference is no longer as strongly advocated as it once was: animals are probably capable, at least to some extent, of symbolic use of language (e.g. Spiegel.de); animals are probably capable of more complex thought operations (e.g. Spiegel.de); some animals have the capacity for self-reflection (Spiegel test); some animals are capable of passing on acquired knowledge to their own offspring, so that animals also develop cultures in the broadest sense. The bathing snow monkeys in Japan, for example, have become famous; some animals apparently develop simple ideas of justice and injustice and fairness (research by Sarah Brosnan and Frans de Waal with capuchin monkeys; also on Youtube). All data according to: https://www.brgdomath.com/philosophie/anthropologie-tk11/mensch-und-oder-tier. (retrieved 04/07/2018).

comparison to the animal, with regard to his abilities and dispositions, and comes
to the conclusion: "… that man is a middle creature among the animals, i.e. the
elaborated form in which the traits of all species gather around him in the finest
epitome" (Herder 1784, p. 59). The instincts were not stolen from man, but sup-
pressed and "ordered under the rule of the nerves and the finer senses". This is at
the same time linked to the idea of development: "The human child comes into the
world weaker than any of the animals; obviously because it is formed to a proportion that could not be formed in the womb (…). Man alone remains weak for a long
time (…) he had to (…) come into the world weak <u>in order</u> to <u>learn reason</u>" (Herder
1784, pp. 114 and 115).

This idea of too early birth will recur with Adolf Portmann in the terms "habitual early birth" and "extrauterine spring" (Portmann 1969) and it plays a central
role for the argumentation of the ontogenesis of man also today. Man, capable of
learning, must learn because he can do less by nature; by refinement and distribution of his powers he has received new effective means, more and finer tools – what
he lacks in intensity of drive he has had replaced by "spread and finer coordination". The animal is born into a narrow world of its own, in which it remains im-
prisoned, but because it is bound to this one world, it moves there all the more
surely and artfully: "The caiman and the hummingbird, the condor and the pipa,
what do they have in common? and each is organized for its element, each lives and
weaves in its element (…) <u>each creature therefore has its own, a new world</u>"
(Herder 1784, p. 70).[4]

Man, on the other hand, does not have such a narrow and limited world. His
senses are open, his organization unspecialized. He is weaker than any animal, but
"organized to finer senses, to art, and to speech." "It is said of one another that man
is without instinct, and that this lack of instinct constitutes the character of his race;
he has all the instincts that a terrestrial animal possesses around him; only he has
tempered them all in his organization to a finer proportion" (Herder 1784, p. 113).

Instead, reason is inherent in man; however, not as a substitute for instinct, as a
faculty that he simply has: "Theoretically and practically, reason is nothing but
something <u>conceived</u>, a learned proportion and direction of ideas and forces, to
which man has been formed according to his organization and way of life" (Herder
1784, p. 115). Consequently, according to Johann G. Herder, man is also organized
for freedom; a freedom that compels him and at the same time enables him to find
"substitutes arising from the midst of his defects." "The animal is only a stooping

---

[4] The underlining that occurs in the text can thus be found in the original text.

slave," the "man is the first freed of creation: he stands upright." To look far about with his head erect (ἄνθρωπος), and thus also to see much that is dark and wrong, is peculiar to him; in this idea lies the whole ambivalence of man: Though he is "set free" in the sense of being liberated from a narrow, instinctual world, yet this very freedom makes his existence uncertain, and what the animal always already has in the way of dispositions to endure in the world that suits him, man must always first acquire as reason,[5] exposed to failure. "Thus, in error and in truth, in falling and rising again, man is a weak child, but still a freeborn one" (Herder 1784, p. 117).

Literally all the great motifs of future philosophical anthropology are already contained in these considerations: the instinctive weakness of man in comparison with the animal; organic unspecialization; the needy existence in an open world; the necessity of learning from childhood (work, which becomes a condition of existence and a moral duty with him, and which later and no longer idealistically understood should be interpreted as the necessity of reproduction); the too early birth and the role of language; finally, the social character of the activity in which man creates man. Language functions as a primary source of knowledge and already in perception man is "metaschematized", anticipating an insight of Gestalt psychology and a preliminary category of Jean Piaget. Contemporary "pedagogical anthropology" is also familiar with these ideas. Moreover, Johann G. Herder's "Abhandlung über den Ursprung der Sprache" (Treatise on the Origin of Language), which was awarded a prize by the Academy of Sciences in Berlin in 1770, is a treasure trove of ideas that has been exploited by later scholars (similar to the writings of Georg Simmel), in which his idea of human history as a reason-guided continuation of natural history is formulated. Two insights are to be noted. What appears as deficiencies in man (in comparison to the animals), as a deficit of a specialization, is not settled by Johann G. Herder as a flat fact; it is transformed by his negation into a new quality, into man's determination to reason. To insist solely on the defect idea would be to grossly misunderstand him. The second thought, which cannot be emphasized clearly enough, stresses the dependence of human beings on one another as a principle of the history of humanity, which he described thus: "As much as man imagines that he can bring forth everything from himself, he nevertheless depends on others for the development of his abilities" (Herder

---

[5] Of course, this thought is not new. Already Blaise Pascal spoke as follows: "Man is only a reed, the weakest of nature; but he is a thinking reed" (Pascal o. J., p. 61). David Hume took a much more sociological view of the situation, assuming that man overcame his weakness "through socialization" ("Treatise on Human Understanding", cf. Streminger 1994, p. 219).

1784, p. 264). I will deal with this attachment to the other in ontogenesis in detail later. Some of the motifs mentioned will recur in almost striking form in Arnold Gehlen's anthropology, but also in Max Scheler's.

1.1.2 Max Scheler: "The Position of Man in the Cosmos"

Many consider Max Scheler (1874–1928) to be the actual founder of philosophical anthropology. In any case, however, it can be demonstrated with him as an example that philosophical anthropology is understood as a science that, in the course of modern thought, began to inquire more intensively into the "nature of man", the less clearly the modern sciences were able to grasp the "role of man" in society. It was probably in this sense that Max Scheler's treatise on the "Position of Man in the Cosmos" (1928/1988) and his essay "Man and History" (1929) brought to the attention of contemporary thought the goal of a consideration of man that differs from a merely scientific or merely historical one and can be an anthropology in the philosophical sense. After the turn of the century, the individual sciences had advanced far enough that Max Scheler was able to take up some of the aforementioned motifs again under a changed perspective.

According to him, man occupies a special position in the world in the midst of all living beings, which he characterizes as "worldliness". In an act of distancing, man makes the "environment" into the "world"; this idea culminates also later still in Arnold Gehlen's concept of the "hiatus". In contrast to a species-specific and to a certain extent unbreakable environmental enclosure in which animals live, organized along the correspondence of a few signals and the instinctive reaction schemes that correspond to them, man moves in an environmentally free way. Man can detach himself from the dictates of drive and instinct, he can transcend what is immediately next, what is most urgent, what is immediately relevant to life: he "transcends." In this respect, Max Scheler also calls man a no-sayer, an ascetic of life (Scheler 1988, p. 55). (In psychoanalytic theory, this idea is encountered in the figure of delay of drive satisfaction and the culture-generating effect of sublimation. Sigmund Freud: "Beyond the Pleasure Principle.") As the basis of this "constitutional No to the drive" Max Scheler introduces – the "spirit".

This spirit is a metaphysical construction; it alone coordinates, directs impulses – not unlike David Hume's soul. Although the spirit derives all power and activity from the urge to live, it alone can direct these impulses; to the blind and empty urges it gives ideas and goals to be realized. Historically the spirit was originally impotent and the urge blind; but through constant interpenetration there comes a spiritualization of the urge of life and a vivification of the spirit. The pure spirit (pure divinity) is seen by Max Scheler in the contour of the world-historical process on the way to its realization (the proximity to Georg W. F. Hegel is unmistakable here).

> "In any case, man – in relation to the animal, whose existence is the embodied philistinism – is the eternal 'Faust', the bestia cupidissima rerum novarum, never calming down with the reality surrounding him, always eager to break through the barriers of his now-here-so-being, always striving to transcend the reality that surrounds him – including his own respective self-reality. In this sense, Sigmund Freud also sees in man the 'drive repressor'. And only because he is this – through this not occasional but constitutional 'no' to the instinct – can man superstruct his world of perception through an ideal realm of thought, and on the other hand, through this very thing, supply his indwelling spirit with the energy slumbering in the repressed instincts. That is, man can 'sublimate' his drive energy into spiritual activity." (Scheler 1988, p. 56)

As much as Max Scheler seems to fall behind Johann G. Herder by this metaphysical justification of the special position of man, he on the other hand empirically argued the motif of cosmopolitanism and environmental-world distance, based on the investigations of Jakob v. Uexküll (1864–1944). Although this by no means created a basis for a subject-society theory, it seems rather to have undermined it, the question of the justifiability of a special position of man in the world was nevertheless driven further – and if it was done by a kind of negative definition: If the mind cannot be conceived as a "deus ex machina" in a biologically desperate situation as a substitute for organ deficiencies – what is it then that philosophical anthropology is searching for? A piece of another answer was provided by Helmuth Plessner.

1.1.3   Helmuth Plessner: "The Stages of the Organic and Man"

The book with the same title was published in the same year as Max Scheler's "Man in the Cosmos"; it frees anthropology from its metaphysical entanglement, but at the same time also from the idealistic philosophy that characterized Johann G. Herder. The Christian body-soul scheme no longer applies, nor does the Cartesian scheme of body and mind – anthropology becomes doubly neutral, as Jürgen Habermas has called it: it becomes neutral with respect to the old Aristotelian idea of a stage organization of the living in the sense that the lower stage form is superimposed by the next higher one; but it also becomes neutral with respect to a Judeo-Christian psycho-physical dualism.

Plants, animals, humans are now analysed in relation to their "sphere", to their environment and world; the relationship in which body and environment stand to each other, the positional form, becomes the starting point of anthropology. This position is not won by Helmuth Plessner in a simple way. In many cases, the argumentation moves on the heights of abstraction of a hermeneutics constituted as philosophical anthropology, which, for its part, is carried out as natural philosophy or phenomenology, with reference back to the modern natural sciences. He never writes in concrete, fact-saturated terms, with concessions to ordinary thought. This may have contributed to the fact that Helmuth Plessner's conception has been unduly little received and processed, and that sociologists in particular have rarely taken up his wealth of thought on the subject of the relationship between man and environment, inside and outside (cf. also Plessner 1985, X, p. 321 ff.). Starting from the Cartesian division into res extensa (measurable external world/body) and res cogitans (inwardness-thinking, feeling, willing/soul), he tries to redefine the relation between thing and environment; the concept of "boundary" becomes the guiding idea, because "the phenomenon of aliveness is based only on the special relation of a body to its boundary." The particular idea that Helmuth Plessner elaborates can be roughly stated as follows: In the sensory perception we have of representational phenomena, boundary is form-boundary as shape and contour; but beyond that, boundary is also aspect-boundary in the sense of whether it delimits a body or constitutes a transition between it and its environment. Thus, if life is really based on a "skin-like" relation of the body to its boundary, of the mass of a thing to its form, of matter to its shape, of "filling" to its "edges," the difference between organic and inorganic bodies "will be phenomenally graspable and will not coincide with empirical differences" (Plessner 1975, p. 123). The inanimate body "is as far as it goes-where and when it comes to an end, its being also ceases." It is different with the living body. Its limits not only include it, but just as much relate it to (connect it with) the medium, and to that extent it is "beyond it." Boundary is a twofold transition.

What does it actually mean, "a body is beyond it when it is measurably there and there at an end, or it is towards it when it demonstrably bristles with solid being up to its boundary contours, up to the edge"? (Plessner 1975, p. 128). A living being appears set against its environment. From it, the relation to the field in which it is, and in the opposite sense, the relation to it goes back: double aspectivity. The inanimate body is free from this complication. It breaks off at its boundary = limitation. "In its liveness, then, the organic body is distinguished from the inorganic by its positional character or positionality (…). As a bearer of the boundary at the same time between and bridging the between, it separates the foreign zone from the proper zone, in order to connect both zones with each other in it" (Plessner 1975, p. 196). The proper zone "disintegrates into itself," as Helmuth Plessner calls it; it cannot be an independent identity, it must become mediated; only in this way does it maintain the connection with the foreign zone. In connection with the above quotation about construction and deconstruction, the essential trait is thus marked: the "double meaning of <u>autonomous self-transformation</u>" (my emphasis).

The self-preservation of the living individual is thus based on an antagonism between assimilatory and dissimilatory processes, and thus does not coincide with either of them (the similarity between this idea of Helmuth Plessner and Jean Piaget's concept of biological-cognitive adaptation is striking).

With this figure of thought, Helmuth Plessner provides a foundation in philosophical anthropology for what in sociology is called the mediating relationship between the objective and the subjective. If an organism were to relate to its environment like any body, the relations between it and the medium would have to run reversibly in the opposite direction. It would interact with everything and everything would interact with it. The maximum of viability would coincide with a maximum of fitting-in, "inserted like the metal core into the casting mould". That such a view of the relation between organism and environment is untenable is shown by the many inconsistencies in nature. The organism would be a plaything of forces, and life would consist in the ceaseless striving to adapt in order to reduce disturbances. But it is a matter of self-initiated appropriation and change.

"Both the doctrine of exclusive fittedness and that of exclusive adaptation overlook the fact that life is essentially both (…) because they grasp the relation of life-bearer and medium reversibly in the opposite sense as a simple physical relation between things in space and time, but not (…) as a non-reversible countersensuous relation" (Plessner 1975, p. 202). What then is the basis of the "double meaning of autonomous self-transformation"? The organism harmonizes in substance and form within certain limits with the medium, without entering into an absolute bond through this harmony; it must fit into the medium and at the same time have leeway. In other words, it is part of the content of the field of position and is at the same time always the centre and the periphery of it.

Conformity and adaptation are realized at the same time in every living act. "It is, as can be seen, the pendant case to the law of assimilation and dissimilation. There the living being bridges the essential abyss between itself and the other things by shifting the border antagonism, as it were, inward into its own central fullness of being and by opening its borders with the antagonism of the circular processes of building up and breaking down for the inflow and outflow of substances, energies, for its incorporation, that is, into the context of things" (Plessner 1975, p. 204). Adaptation and conformity as simultaneously realized in every living act point to another problem: the organizational form of the relation between organism and field of position. "For the living thing there is here a radical conflict between the compulsion to be closed off as a physical body and the compulsion to be open as an organism" (Plessner 1975, p. 218).

Because the naturalness of animals is inaccessible to him, obscured by eccentricity, he is artificial by nature in his form of existence. Man lives only by leading a life. Here a thought runs from Johann G. Herder and Johann G. Fichte to Helmuth Plessner and Arnold Gehlen. According to Helmuth Plessner, the eccentric form of life and the need for supplementation thus form one and the same fact – therein lies the movens for all specifically human activity, "the ultimate reason for the tool and that which it serves: culture" (311).

This somewhat longer presentation of Plessner's considerations was important to me for two reasons: First, they hold a fruitful starting point for an individual-environment theory whose foundation is laid on a biological-anthropological level with the acquisition of action competence, thought, and language (see below), thereby largely saving it from metaphysical justifications, and, moreover, it allows one to think of the justification of a social theory as a cultural theory; on the other hand, from this biological-anthropological level, a whole series of other concepts, which in turn are used for subject-society theories, can be viewed critically (drive-repression model, deficiency-being model, etc.). However, it must be taken into account that the phenomenal determination of conceptuality – especially eccentric positionality – cannot be taken out of phenomenology and brought into or translated into an evolutionary perspective without methodological difficulties (cf. Dux 2017, p. 74). Helmuth Plessner assigns knowledge a central position in human beings, from which it is evident that it is not innate, nor is it predetermined by their mental existence (here Johann G. Herder shines in again), but is connected to the anthropological constellation (see further below in Günter Dux) and the competence to act that emerges from it. For the question of participation, which will still concern us in detail, it should already be noted here: – the evolutionary perspective – that although the acquisition of action competence takes place in the early ontogeny of subsequent generic members, it would not have been possible if conditions of interaction and communication with social others had not arisen in ontogeny (Dux 2017, p. 78). This can also be seen in Arnold Gehlen's methodologically different considerations.

1.1.4   Arnold Gehlen: Deficiency Being and Institutional Theory

That it will no longer be necessary in the future to speak of man as a spirit-creature created by God, nor of man as an "arriviste ape", who differs from the anthropoids essentially only in his specific intelligence, was demonstrated by Arnold Gehlen in his book "Man: His Nature and His Position in the World", first published in 1940. For the third edition, the author then heavily revised the last part of the book of the first edition in order to free it from regulatory ideas of a conservative stabilization theory, which were very much in line with National Socialist values and virtues. Despite this ballast, however, Arnold Gehlen's work has become a source of subtle debate for many authors, including critical and left-wing authors (cf. Rehberg 1986).

The old Herderian theme of how man learns to lead his life "from the midst of his deficiencies" is brought into the center of a biological-anthropological analysis in Arnold Gehlen. The perspective is a comparative one between animal and human, the principle of cognition a pervasive structural law "that governs <u>all</u> human functions from the bodily to the spiritual." This structural law becomes visible in the <u>determination of man to act,</u> in the determination of "man as a stance-taking, non-established, disposing being (also disposing of himself)," enforced from his physical organization-or, as Arnold Gehlen puts it in a question, "How can such a defenseless, needy, such an exposed being keep itself alive at all?" (Gehlen 1986a, p. 19). The answer emerges from the matrix: Animal– Environment/Human – World. The "exposed and risked constitution" emerges from this matrix:

"The 'environment' of most animals, and especially of the higher mammals, is the non-interchangeable milieu to which the specialized organ structure of the animal is adapted, within which again the equally species-specific, innate instinctual movements operate. Specialized organ structure and environment are thus concepts that presuppose each other. If now man has <u>world</u>, namely a clear non-limitation of the perceptible to the conditions of biological holding, then this too means at first a negative fact. Man is open to the world means: he <u>lacks</u> the animal fitting into a cut-out milieu. The immense openness to stimuli or impressions in the face of perceptions that have no innate signal function undoubtedly represents a considerable burden that must be coped with in very special acts. Man's physical unspecializedness, his organic destitution, as well as the astonishing lack of genuine instincts, thus form among themselves a connection to which 'world-openness' (Max Scheler) or, which is the same thing, environmental deprivation, form the antithesis" (Gehlen 1986a, p. 35).

But what is the origin of unspecializedness, organic destitution and the lack of instincts? Morphologically, man, in contrast to all higher mammals, is primarily determined by deficiencies; in the exact biological sense, these are "maladaptations," "unspecialized features," "primitivisms". The hair coat is missing and with it the natural protection against the weather; the natural organs of attack, the formation of the body suitable for escape, are missing; in comparison to almost all animals, man has weaker senses and an "almost life-threatening lack of genuine instincts" (Gehlen 1986a, p. 33) – which beyond that, however, represents an almost tremendous difference, rich in consequences: It is a kind of "physiological premature birth," a "secondary nestling," the "only case of this category among vertebrates" (cf. also Portmann 1969). Thus, while unspecialized means that within natural, primordial conditions, humans would have long been extinct as ground-dwelling amidst the most agile flight animals and the most dangerous predators (Gehlen 1986a, p. 33), the fundamentally significant "extrauterine spring" means the following: In it, maturation processes must be combined which as such were already fostered in the womb, but now, because of the too early birth, with the restricting experiences of innumerable stimulus sources and their processing, such as the acquisition of the upright posture, the movement repertoire, and language, are only progressing – so that "a number of ontogenetic peculiarities, such as the duration of pregnancy, the early mass development of our body, the degree of formation at birth can only be meaningfully understood in connection with the mode of formation of our social behavior" (cf. also Portmann 1969). This conception of an organically deficient being, therefore open to the world, i.e. not naturally viable in any particular environment, has two specific implications: On the one hand, at the price of the demise of the species, man is given the task of producing the means for his survival in the first place through self-acting and expedient transformation of the natural world; on the other hand, man is not designed in his affective and cognitive processes for fixed stimulus-instinct schemata. Due to the "plasticity" of the human being, his drives can be bound in many ways; his world is not a cut-out environment, but "open". This results in a serious consequence: Man's openness to the world is a <u>burden</u> for him. However: "The basic idea is that all the 'deficiencies' of the human constitution, which under natural, so to speak animal conditions represent a maximum burden on his ability to live, are turned by man, acting on his own, into the very means of his existence, in which man's determination to act and his incomparable special position are ultimately based" (Gehlen 1986a, p. 37).

A task of physical and vital urgency for man is therefore relief: to rework the deficient conditions of his existence into opportunities for his life's limitation. In this interplay of burden and relief via action, man does something very specific: he copes with the reality around him by <u>changing</u> it into something useful for life, and, from the other perspective, he extracts from himself "a very complicated hierarchy of achievements, 'establishes' in himself a structural order of ability that lies in him merely according to possibility" (Gehlen 1986a, p. 37). In the background of this thought is a "law" of philosophical anthropology that finds the necessity for the formation of a specifically human ability in man's peculiar biological situation: self-acting production. The theoretical category for this is called freedom by Johann G. Herder, positionality by Helmuth Plessner, and by Arnold Gehlen: Action.

Arnold Gehlen immediately follows these considerations with the transition to the concept of culture, for the epitome of nature reworked by man into what is useful for life is called culture. The cultural world is the human world (not the environment!), the "detoxified" nature, the "second nature". The peculiar core of this consideration, which is at the same time strongly reminiscent of Helmuth Plessner, is the complete "unnaturalness" of human existence; the "unnatural" culture is the effect of a unique being in the world, itself unnatural, i.e. constructed in contrast to the animal. In precisely the place where the environment stands in the animal, therefore, stands the world of culture in man. Thus, in man, the unspecializedness corresponds to the openness to the world, and the impecuniousness of the physique corresponds to the culture created by him.

Relief and action are the key categories of Gehlen's conception. From them he understands all social moments, from the habitualization of action to the cultural order of institutions, as relief from the deficiency situation – and in this line of thought lies the core of his socialization theory. The unpredictable field of surprise is narrowed down by making experiences through action. Arnold Gehlen illustrates this with the autonomous movements that fill childhood:

The things

> are seen, touched, moved, treated in communicative movements of contact (…) The success of these processes (…) is that the surrounding world is 'worked through' in the direction of availability and completion: things are pulled into handling and put down one after the other, but in the course of this process they are unnoticedly enriched with a high-grade symbolism, so that at last the eye alone, an effortless sense, overlooks them and finally sees in them values of use and handling which were previously laboriously experienced by itself. (Gehlen 1986a, p. 41)

This results in a specific concept of orienting oneself: It means reducing the flood of impressions to certain centres and freeing oneself from the pressure of the immediate abundance of impressions. The cognitive and motor aspects are thus closely linked. Always, namely up to the highest psychic and cognitive performances, the appropriation of the world is at the same time the appropriation of oneself, the statement to the outside is at the same time one to the inside. Language now ties in with the world of things, which has already been narrowed down to "symbols," and thus frees them from the necessity of being present in reality in order to be able to be processed. Reality thus becomes available and manipulable in language and thought, a new enormous relief. This relief made possible by thought and language is connected with the drive life; it must have a special structure in an acting being, it must be orientable "i.e. contain not only certain vital needs, but also the often very conditional circumstances of their satisfaction, with which, because these themselves change, it must vary along" (Gehlen 1986a, p. 53). This idea is found in Talcott Parsons' concept of "need-dispositions".

The displaceability of the drive plays a decisive role; even the most circumstantial and indirect actions ("e.g. the preparation for the production of means") must still be able to have a driving interest.

> Between the elementary needs and their external fulfillments, changing according to unpredictable and accidental conditions, is interposed the whole system of world-orientation and action, that is, the intermediate world of conscious practice and factual experience, which runs through hand, eye, sense of touch, and speech (....) It is now the same instinct-reduction which, on the one hand, dismantles the direct automatism which, given a sufficient inner stimulus-mirror, when the assigned trigger appears, disinhibits the innate reaction, and, on the other hand, sets in freedom a new system of behaviour relieved of instinct-pressure (....) there is an extensive independence of actions as well as of perceptual and thinking consciousness from one's own elementary needs and drives, or the ability to 'unhook' both sides, so to speak, or to expose a 'hiatus'. (Gehlen 1986a, p. 53)

Here is the point at which, in Arnold Gehlen, the transition from a biological-anthropological theory of man to a cultural theory of society takes place. Independent actions, which from an individual point of view were at least originally only means and then became ends in themselves, now become immediate ends for other individuals, are objectified and become institutions with the character of supra-personal normative orders. In another work (Gehlen 1986b), Arnold Gehlen develops this idea using the example of the division of labor. What appears to be ontogenetic, so to speak, from an individual perspective as a relief of needs, the formation of habits, and the enrichment of motives for action, becomes a mutual prerequisite from the perspective of socialization.

"In the simple case of a society in which a few specialists (blacksmiths, potters, or whatever) produce for everyone, but are fed by them, there arises, anthropologically speaking, the state of mutual relief of need. The need for food of the 'specialists' moves into the state of 'background fulfilment', the certainty of permanent virtual fulfilment" (Gehlen 1986b, p. 33). In the state of division of labor, work now turns into "habit formation saturated with intrinsic value," into which new motives flow from the object of work itself or from social relations, which the worker adopts and invests. This behavior develops away from the immediate next needs, it becomes independent. The interests of others can be attached to these activities, one serves the interests of the other and vice versa. "The structure of production and distribution that thus arises in a circle now also becomes independent objectively, as a process into which the individuals enter and from which they die away, and it becomes independent subjectively, in the consciousness of the participants of the existence of a valid order" (Gehlen 1986b, p. 34).

In institutions, our individual needs "intertwine" with "the general, objective necessities which the existence of society develops." These institutions are functional for man: "All (sic!) institutions of work, rule, family, etc., have today as always a direct fulfillment value for human primary needs, but they become independent vis-à-vis man and one acts from them, in the sense of their preservation, their self-demands, their laws" (Gehlen 1986b, p. 18). Moreover, they fulfil functions of relief from subjective motivation and the constant improvisation of decisions to be represented on a case-by-case basis. The process of the emergence of institutions goes, as it were, from the transformation of work (manufacturing) into a habit formation saturated with intrinsic value (i.e., the making independent, habitualization of groups of motives and action sequences), via a then possible influx of new motives stimulated by the object itself or by social experience (i.e., the virtual shift of purpose through newly added purposes), to the path development of this behavior from the immediately proximate needs and to the linking to the interests of others. Anthropologically, the subject of habit formation, the completely unproblematic experience of performing an action (in which all reflection is suspended, because one cannot act and reflect at the same time, but can only "watch" one's actions), is of central importance: all institutions are lived as systems of distributed habits. In this understanding, habitualized action in institutions also has the purely actual effect of "suspending the question of meaning. Those who raise the question of meaning have either lost their way or are consciously or unconsciously expressing a need for institutions other than the existing ones" (Gehlen 1986b, p. 61). There is a central answer to Gehlen's question of how it is possible for an instinct-bound, yet drive-surplus, environmentally free and cosmopolitan being to stabilize its existence: by creating institutions.

Now, however, institutions acquire a peculiar quality in Arnold Gehlen's work; in them, man does not appear organized for freedom, as in Johann G. Herder, but for suppression and regulation, for repression. The concept of action, which is in itself broad and open to meaning, is primarily restricted to the task of learning to control behavior through mutualization in the face of the surplus of plastic drives, the plasticity and variability of behavior, the sensory overload; furthermore, the same institutions that people produce in their mutual thought and action become an independent power "that in turn asserts its own laws right into their hearts" (Gehlen 1986b, p. 8); man must "allow himself to be consumed" by these historically evolved realities, and this is because the autonomy and self-sufficiency that institutions gain vis-à-vis the individual are to be derived from man's nature. While Jürgen Habermas, in his critique (Habermas 1970), focuses on the fact that Arnold Gehlen elevates the "breeding and harshness of archaic institutions" to a historical invariant, seeks anthropological constants, and thus makes "blind domination" the normal case, i.e., fails to recognize the variability of conditions under which a human being could one day emerge who is independent of large "breeding systems," I see the point of departure as having been shifted backward, as it were, another stage: In that Arnold Gehlen fixes human nature by a biological-anthropological rationale, and the solution to the deficiency dilemma can only ever be learning to control behavior, no path at all opens to institutions that cannot become repressive; even in the thought of historicizing institutions, this conclusion returns, treacherously in language: "Culture is by its very nature a working out, over centuries, of high thoughts and decisions, but also a recasting of these contents into fixed forms, so that they can now be passed on, indifferent to the small capacity of small souls, to survive not only time but also men" (Gehlen 1986b, p. 24).

The methodological weakness lies in the idea of wanting to derive a certain model of society from anthropological basic facts (constants); this seems to me to be a legacy of the older social anthropology. However, Max Weber already stated that recourse to the most general and therefore content-empty laws is "worthless" for an analysis of social phenomena. Almost fifty years ago, therefore, an authoritative judgement was: an anthropology, as it exists today, is at best capable of justifying the phenomenon of man's self-activity and the compulsion to act; social criticism, for its part, must be socially justified in such a situation. Otherwise, the conclusion is inescapable that man can only realize his anthropological destiny in institutions (cf. Lepenies 1971, p. 82).

Moreover, it should be noted: The conditions of human existence, which are probably coherently employed in Arnold Gehlen's conception: openness to the world, plasticity and surplus of drive, not being fixed and being at risk, etc., are thematized one-sidedly from one point of view – that of relief. Johann G. Herder, it seems to me, solved this problem more "elegantly". That the condition presupposed here, the experience of the burden of this situation, would actually be demonstrable and not merely postulated, is not shown in Arnold Gehlen (cf. e.g. Geulen 1989, pp. 50 and 51). Another objection hits the postulate of the universal functionality of institutions; their functionality is related to human needs, subsequently to relief services, and thus remains on an anthropological-individualistic level (as once in Bronislaw Malinowski; [cf. Amann 1996]). Such a notion of functionality is quite untenable because it does not allow for a methodological-theoretical approach to society as a supra-individual reality sui generis. In the course of the discussion about the "fundamental change" that is taking place today, which decuvrates progress of the conventional kind as a step backwards and raises the question of the saving alternative (Günther Altner), the anthropological conception of man also came under criticism. Thus Günther Altner (1987), while at the same time paying tribute to Johann G. Herder, already criticized Arnold Gehlen in particular for his nature-despising concept of man. "Arnold Gehlen, for example, corrupted the special position of man meant by Herder by letting the alleged biological deficiency of human nature become a compensatory challenge for the cultural being man" (Altner 1987, p. 29).

The specific position of man, which forces him to change nature in a life-serving way, does not allow him to be understood as a spiritual being, but only as a nature-changing cultural being (however, also Günther Altner again drew in a metaphysical concept of spirit). Nature is degraded to a means, to an instrument and at the same time an object, whose transformation only brings forth that which is primarily appropriate to man, the "second nature", the "culture"; treacherous, indeed, also the already mentioned thought in Arnold Gehlen that the culturally untouched nature is the still "non-detoxified" nature. Such anthropology, said Günther Altner, "reads like a legitimization of that ever-deepening gap between man and nature that we encounter today in the eco-crisis" (Altner 1987, p. 32). However, the criticism, which is certainly at the right point here, then becomes blurred when Günther Altner thinks that the "forgetfulness of nature" that is shown here in a completely undisguised way is characteristic of the anthropological tradition in Europe. Seen in this way, it would become a general cultural critique that misses the specifics of social reality. The subjugation of nature, in terms of its form and content, has always been tied to the concrete economy (the economy is the natural ground of logic from the very beginning, said Joseph Schumpeter); it seems to have been "problem-free" as long as it did not become exploitation. But this point cannot be fixed historically, it can at best be named in myth: when the cry rang out: "The Great Pan is dead". Already in the first appearance of Antigone (Sophocles) the chorus speaks in the trembling praise of man of how the rape of nature and the civilization of man go hand in hand.

The critique of a rape of nature therefore seems to have a deeper starting point: not in the dominance of images of man (from René Descartes to Arnold Gehlen), but in the interplay of production and its corresponding ideology. It was not until economic production had become unlimited overproduction and theories of man provided it with ideological legitimation that the fate of nature was finally sealed. Here, however, the idea is correct again: no tradition since the eighteenth century in Europe has contributed as much as that of the image of man to suppressing critical questions that wanted to doubt man as Prometheus, as the all-maker and all-capable.

1.2   Günter Dux: The Epistemological Turn

If we take, for example, an anthology on philosophical anthropology from 1975, which at that time was understood to be at the height of the times (Gadamer and Vogler 1975, 2 vols.), it is striking that in numerous individual contributions for the determination of the specifically human, from language to cognition and action and from faith to understanding, argumentations on the theory of reasons occur, which were obviously nourished from older historical-philosophical and metaphysical stocks and nevertheless claimed to be contemporary. Nothing could be a more suitable starting point for Günter Dux's epistemological reflections than a view of the world founded by such means.

The human form of life, this thought (already anticipated in Greek antiquity) runs through the Enlightenment and anthropology, is a spiritual form of life. First and foremost, it expresses itself in action, thought and language, and in a systemically connected way. However, the question of how this spiritual form of life became possible and could be created by man himself has hardly ever been posed and answered in a way that is suitable for epistemological criticism. It would mean to take full account of the fact that man's form of life belongs to a thoroughly secularly understood universe and has formed itself out of evolution, whereby spirit can no longer be traced back to a higher (or lower) principle that somehow lies outside ourselves, as has happened since antiquity and is still sometimes attempted today. A sociologically understood theory of man and society must, under this presupposition, assure itself of the anthropological or evolutionary foundations without which the emergence of the species Homo sapiens is inconceivable, and it must at the same time seek an explanation for the social emergence of this form of life. For centuries, "spirit" was sought at the bottom of the universe, as always already given, not capable of being left behind, or anchored in the individual as a faculty and in this faculty also taken out of nature, precisely the absolute at the bottom of the world or before all the world – as "being in the world without being of the world" (Dux 2017, pp. 11 and 12). God was probably most often claimed for this. At present, some research still finds that the spiritual form of life, action, thought, language, can be located in the genome and, mediated by the genome, in the brain.

Could this be a questionable remnant of the argument from the point of view of epistemology? The brain does not think, it is in its central constitution an organ, and Heinz v. Foerster says its vocabulary is "click, click." According to Günter Dux, it is important to move away from a way of thinking that seeks absolutes and to acknowledge "that with Homo sapiens, who emerged from evolution, his social competencies and, in one sense, his cognitive competencies were also able to develop further historically" (Dux 2017, p. 17). The formation of this form of spirituality as we know it today is the result of a historical development that occurred with the emergence of the cultural life form of Homo sapiens between 140,000 and 40,000 years BC. Where is the starting point of a possible reconstruction to be found? It lies in what Günter Dux calls the "anthropological constellation" of humans (Dux 2017, p. 63). In the transitional field between hominids and hominins[6] the brain grows and in the course of this development three structure-forming possibilities open up for humans: the opening of the world, the dwindling of organic circuits (instinct dependence) and the constructive construction of the world (Dux 2017, p. 37). The organism could respond to this development in only one way: by forming a mental-constructive form of life (only in its development could it remain in the world and at the same time distance itself from it). It takes place through the development of agency, language, and thought in interdependence. Only to these structural moments or strategies is the whole potential inherent from which family, community, society, religion, power, domination, etc. were created. Action competence allows the changing intervention in the world, language allows communication and, in the way of representation of action and objects and events in the world, thinking (also thinking about thinking) – none without the other.

Günter Dux's reasoning starts from the so-called Copernican turn in Immanuel Kant, in which the strategy of cognition is reversed and the demand arises to let objects be guided by our cognition and not vice versa. The idea is formulated in the well-known sentence: "The understanding does not draw its laws (…) from nature, but prescribes them to it."[7] I. e. nothing else than that we ourselves at least partly

---

[6] Hominins is the name given to a subfamily of the great ape family (hominids). This subfamily includes the species of the genus Homo, including humans living today (Homo sapiens), and the extinct ancestors of this genus, but not the common ancestors of chimpanzees and Homo. The only non-extinct species of hominins is humans. Belonging to the hominins is called hominin, belonging to the hominids is called hominid (according to Bernard Wood (2011) *Wiley-Blackwell Encyclopedia of Human Evolution*). Günter Dux uses the term hominins to designate those creatures that also evolved culturally towards Homo sapiens during the two million years of the Pleistocene.

[7] It is not unimportant to see that Karl Popper, in the first volume of The Open Society and Its Enemies (first published in 1945), quotes this sentence and calls its formulation "brilliant" (Popper 1970, p. 15).

generate that order which we find in the world. It is we who create our knowledge of the world, it is we who construct those orders which we then think we find in the world as orders. This refers to what Günter Dux calls the "convergence theorem", according to which all derivation of knowledge converges on the individual. Immanuel Kant's blind spot lies in letting the faculty of cognition be innate to the individual; the way to the conception of a constructive understanding of life forms is still far. It is only in the twentieth century that (e.g. with Jean Piaget) the epistemological strategy is echoed: to gain the origin of the mental life form by reconstructing it from the ontogenesis of the descendant species members (the "theorem of constructivity"). These two moments culminate in the "theorem of historicity": if the world is a world of man's own making, then this must always have been so and must be reconstructible as the history of the human form of life (Dux 2017, pp. 19–21). If all this is true, then man's spiritual form of life is "self-determined." Self-determination represents, in a sense, the armature in the understanding of man as gained in the early modern period (Gerhardt 1999). The anthropological argument is that the evolution of the brain laid the foundations "that the organizational forms of man's conduct of life could be created by man himself." The ontological argument determines that the "basis of the organizational forms of the mind: action, thought and language, (…) must be trained anew on the world by each descendant species member as a practice form of the conduct of life." The social argument holds that "human forms of life are created by the systemic interconnection of subjects in society." For this, communication and interaction between the members of society are constitutive, which also makes it clear that action, thought, and language are socially constituted and thus to be reconstructed historically and empirically (Dux 2017, pp. 25 and 26), in inescapable referentiality of people to one another.

1.3   Relation of Participation to Evolution

Just as in the 1970s and 1980s, when sociobiology gained strength and set out to explain social behaviour patterns among humans from biological evolution and heredity, so we are also experiencing today – at least I cannot completely escape the impression – a boom in argumentations that now claim, for example, that altruistic and cooperative behaviour or processes of understanding are innate, or that human thinking is only gradually different from that of the apes.[8] Despite all the

---

[8] I will not go into the already unmanageable entries on the Internet. From the scientific literature in book form, reference should be made, for example, to an author who now enjoys considerable presence in the media: Michael Tomasello (2013).

methodological rigour put forward in many studies, the question remains open as to whether experimental results from a different epistemological perspective would not also permit other interpretations and thus different findings. Bearing in mind the remarks in Sect. 1.2, in which I dealt with Günter Dux's work in more detail, what is still at stake here is the basic idea of a world produced by man himself, which is based on the evolutionary substrate of Homo Sapiens, but which is no longer controlled by this biological basis (perhaps with the exception of some so-called instinctual remnants). I consider this starting point important because it seems to contain all the possibilities that could lead to the historical-cultural forms of human social behavior that have actually emerged and in which social participation is inscribed. For the sake of experiment, I take the view that the decisive question for sociology is not that of instincts and drives, concepts that are charged with shimmering meanings and misunderstandings as a result of their history, but rather that of the conditions under which human beings could and had to constructively create their own world.[9]

There seems to be unanimity in the discussions that the evolutionarily decisive step that had to lead to the specific shaping of the life forms of Homo Sapiens was the enormous enlargement of the brain (chimpanzee 400 cubic centimetres, Homo Sapiens 1400 cubic centimetres) and, parallel to this, the anthropological foundation of the human life form.[10] This process harbored three immense upheavals already mentioned: (a) the opening of the world, (b) the dwindling of the organic circuits of behavior, and (c) the constructive construction of the life-world (Dux 2017, p. 37). Historically, these developments (which include new forms of thought and communication) fall within the period that began about 100,000 years ago and was (tentatively) completed about 30,000 years ago. This process is also referred to as the "cognitive revolution", with which myths, gods and religions presumably first appeared (Harari 2015, p. 37).[11]

---

[9] This point of view has nothing to do with apodictics; it remains true: "For since we are the results of previous generations, we are also the results of their aberrations, passions and errors, indeed crimes; it is not possible to separate oneself entirely from this chain" (Friedrich Nietzsche in his discussion of the *critical* way of seeing the past 1980, pp. 229 and 230).

[10] I will not go into the evolutionary features of the upright gait, the opposing position of the thumb on the hand, the considerable energy requirements of the human brain, etc. For the history of the idea of the upright gait, cf. the impressive work by Kurt Bayertz (2012).

[11] Yuval N. Harari gives different time periods in his book; the relative indeterminacy could only be circumvented if specific developmental steps were singled out in each case – and even then only rarely, because new finds regularly force re-dating, such as the one that Homo erectus probably left Africa 300,000 years earlier than previously assumed (daily newspaper "Der Standard" of 12 July 2018). Who would want to make decisions as a layman?

Not only these, but also the earliest linguistic, technical and artistic-creative expressions of humans (hand axes, cave paintings, bone carvings, etc.) point to a connection that has already been mentioned several times: the anthropological situation of humans, the evolutionary "hiatus" (cf. also Helmuth Plessner and Arnold Gehlen) that pushes itself between humans (in the biological sense) and the world, requires action. A living being (no longer a creature!) that no longer has any genetically fixed forms of behaviour in dealing with its external world can only arrive at a competent form of organisation of its behaviour that can take account of this specific circumstance in one single way: it must develop a reflexive relationship to itself and its external world and make its behaviour controllable (Dux 1992, p. 72). It is a matter of the process in which this controllability is acquired (ontogenetically) and formed in the structure of action. This is not to be discussed further here; it has been done sufficiently in Günter Dux, Jean Piaget, etc. More significant for the present topic is the question of how the life interests of the acting subject can become objective and be pursued intentionally in action.

Since the studies of Humberto R. Maturana and Francisco Varela, it has been scientifically or neurobiologically substantiated that the human organism can be considered a self-referential system (e.g. Maturana and Varela 1987), which means nothing else than: Everything that happens in the organism happens in such a way that its states ensure homeostasis. In other words: In the organism there are no other determinants than that of making the connection to past processes in a way that makes the presently occurring processes connectable to future ones, i.e., this recursivity constitutes the strategy to sustain life (Dux 1992, p. 74). This brings us close to the question of what life might mean. In the language of Humberto R. Maturana and Francisco Varela, this means that "living beings are characterized by the fact that they – literally – continuously generate themselves" (Maturana and Varela 1987, p. 50) – in autopoietic organization. Related to this conception is the idea that no human action at all can be free of the biological constitution, whereby the recursivity inherent in the self-referential system translates into self-care – concern for oneself – and the need to seek the means to enforce it. Both components presuppose consciousness and will.

Now the specific connection between evolution and participation is to be considered, which is necessarily based on the anthropological constellation already mentioned. Action competence, interaction and communication interlink to form the primary form of order of the community. In it, the formative moments of normativity and power are formed. The stability of the organisational form is created through the expectations associated with all action. But where do expectations come from? The basic constitution of natural history, out of which the sociality of the form of life develops, shows an inescapable bond to another (mother, reference person, etc.). In the acquisition of competences is determined what connects each one with each other. Everyone acquires "in his ontogenesis also the relation to another as a condition of his own existence" (Dux 2017, p. 182). This means, after acquiring competencies, that the agent is confronted with the actions of others. The point of this is that the agent takes up the position that is characteristic of him or her vis-à-vis the others in the field of action. Action, however, is always confronted with the modal form of possibility: The world is as it is, but it can also be changed, and the other agents can act in this way, but also differently. "Under the modal form of possibility, expectation becomes the basic categorial structure not only in dealing with nature, but precisely also in dealing with others in the social world" (Dux 2017, p. 194). This touches on an old moment of all action theory since the nineteenth century: the expectation of action. Expectations are initially always cognitive expectations, but it does not stop there, they become linked to interests.[12] In this way they become demands on others to comply with them. They are expressed as normative expectations in the form of the ought. To put it briefly: in the ought there is the expectation directed as a request to others to take account of the interests of action. If it is now taken into account that agents are always connected to other agents, and that therefore different interests always meet, it is logical that, in order to regulate conflicts, patterns of action are formed which are evaluated and established as forms of agreement of action in the community. In this way they can no longer be negated, they are removed from arbitrariness and can claim validity. "The normative moment of validity (…) means precisely this finding: to see oneself, with one's actions in a community, exposed to generally accepted judgments as to what is and what is not permissible in terms of actions and pursuit of interests vis-à-vis others" (Dux 2017, pp. 196 and 197).

Participation in life is anchored in this context, from it receives its impulses, its regulations, and also its freedoms. Only in this way can it be understood that even the completely socially-abstinent person in solitary life is still a part of the community – albeit in a special position.

Interests, then, clearly function as goals of action with direct reference to the agents themselves. Humans, unlike animals, to repeat, are disconnected from direct instinctual guidance and must construct systems of action as a cultural organiza-

---

[12] This is where Sigmund Freud's "Probehandeln" would have its logical place.

tion. Humans cannot help but formulate the satisfaction of their needs in action with others into interests and attempt to achieve them through controlled forms of action. Therefore, needs whose satisfaction takes place under competition and becomes dependent on the actions of others are called interests. Thus, purposive action, although not covering the whole form of human existence, nevertheless moves into a constitutive centre (Dux 1992, Chap. 3) of what is considered a principle: human society builds itself up through actions. It is an organizational structure of actions and not simply an ensemble of subjects. In the processes of action, the life interests of the subjects involved become concrete with action and they are pursued intentionally. The ultimate purpose of action is always the agent himself. However, every pursuit of interests also brings power potentials into play. Power is a social fact and not a natural instinct. It emerges as a cultural form of action, and each potential for power shapes itself among those of others. What power a person processes is determined by his or her involvement in the social organization and its distribution and allocation of opportunities. Therefore, the idea of scopes or margins plays an important role in the conception of standard of living. What becomes visible as contradictions in the disputes and interpretations in our societies is therefore an expression of power relations. Power, although serving as a means to satisfy interests elsewhere, can itself become a goal of action that is based on a need: Power is litigated in order to gain more power; therefore, it is rightly seen as limitless and directed towards ever more power. It resides both in the individual and in the social organization. Today's social conflicts are power struggles for life chances.

So far, it should have become clear that participation in life is centrally anchored in those forms of life that humans have produced themselves in historically varying ways and is not owed to an innate need for sociality or other sources of justification theory. Now, not all that much is known about the life forms of hunter-gatherers, for example, although Homo sapiens lived as a game-herder for most of its history. In research on Stone Age living conditions, we rely almost exclusively on objects that have "survived" such as bones and stone tools. Wood, bamboo and leather have mostly rotted away. Written evidence does not exist. Theories that attempt to provide information about ownership, hierarchies, social relations, and social and cultural techniques of existence of that time stand on shaky ground. Evolutionary biology strives for insights and comes, for example, to the conclusion that today's widespread "obesity" is genetically anchored and goes back to the specific nutritional situation of those times (theory of the "glutton gene"), so that today we showed a strong tendency to eat far more than hunger dictates (Harari

2015, p. 58); other research advances the "primal commune" thesis, claiming that marital infidelity and high divorce rates today are due to our being forced into monogamous marriages and nuclear families against our very nature, the exact opposite of what our ancestors supposedly practiced (Ryan and Jethá 2010). Conversely, other research claims that monogamy and nuclear family structures were prevalent at the time and infers that this is why these forms are still dominant in our societies today (Harari 2015, Chap. 3). The first thesis (Fress-Gen) leads directly into the well-known dispute between Darwinists and Lamarckists, the second (Ur-Kommune) is fatally reminiscent of the self-justifications common among young alternatives in the nineteen-sixties, when they were confronted with moral reproaches because of a permissive lifestyle. In accordance with the epistemological position that has already been formulated, such "findings" are not to be classified as reliable enough; in some cases they seem audacious.

The reference to the hunter-gatherer context should once again emphasize the argument that we have only scanty knowledge about the early times of Homo sapiens, which still applies in a similar way to the phase of agrarian existence, as far as his social forms of behavior and networks, his cultural institutions, his power strategies, etc. are concerned. It therefore seems advisable to me to examine the idea of participation in this framework empirically only sporadically, but instead to attempt a conceptual-logical reconstruction with reference to the previous insights.

> *Self-care, which is transformed into interests in the pursuit of the satisfaction of needs in contact and competition with others, must be seen as the fulcrum from which action crystallizes and in which participation is anchored.*

Self-care has entered into the structure of action. From this action, stable forms of behaviour necessarily form, into which competences of action, thinking about the world (its realities and events) and its expression in language have already found their way. These cultural "stocks" then confront the individual quite in the sense of social facts (Èmile Durkheim), which represent a reality sui generis. The notion that the agent makes what he wants the goal of his action requires that he addresses this as an expectation to the other and underpins it with power potentials. The request in performative-perlocutionary form to do something or to refrain from doing something therefore forms the basic structure of the norm. "The normative constitution of a society is therefore always the reflection of those interests and power potentials that have been able to form in society" (Dux 1992, p. 85).

Participation in life can therefore, a first approximation, be regarded as a more or less successful involvement and disengagement[13] in the process of the struggle for interests in society, which is ubiquitous. This involvement/ disengagement is always a balancing work to be done between becoming-included and actively including oneself or becoming-disengaged and actively disengaging oneself. Both are vital, so that one could also speak of vital inclusion and vital disengagement. In both cases, the environment and the willing individual are addressed, for both are indispensably involved in the processes in a reciprocal effect.

The aforementioned "more or less" is determined according to the ontologically acquired potentials and resources on the one hand, and according to the culturally institutionalised standardisations and power structures on the other, framed in the notion of integration into society. The extent and form of integration oscillates across the life cycle and across times in history. If time is now taken into account,

a second approach, participation in life becomes a form of behaviour which, due to social change, historically presents itself as extremely variable and individually (in the life cycle) as a task and cannot be realised at will arbitrarily.

Normative regulations can exclude entire groups of people from a specific form of participation (minors from legal transactions, the older people from working life, etc.) or force them to do so (entry into an insurance relationship, adherence to religious rituals, etc.). Individually unsuccessful involvements or unsuccessful disengagements can have negative consequences for the life course. Similarly, failure to integrate larger groups results in significant changes in the structures of the environment. For all further considerations in this work, participation in life should therefore mean (Fig. 1.1):

---

[13] For the conceptual pair involvement/disengagement, a terminological proximity to that in Erik Erikson: involvement/evolvement or to Anthony Giddens' embedding/disembedding can be assumed. Conceptually, it has nothing to do with either of these, since in the first case the structural conditions are virtually disregarded, while in the second case the constructive self-activity of the subject remains undervalued (cf. Erikson et al. 1986; Giddens 1995).

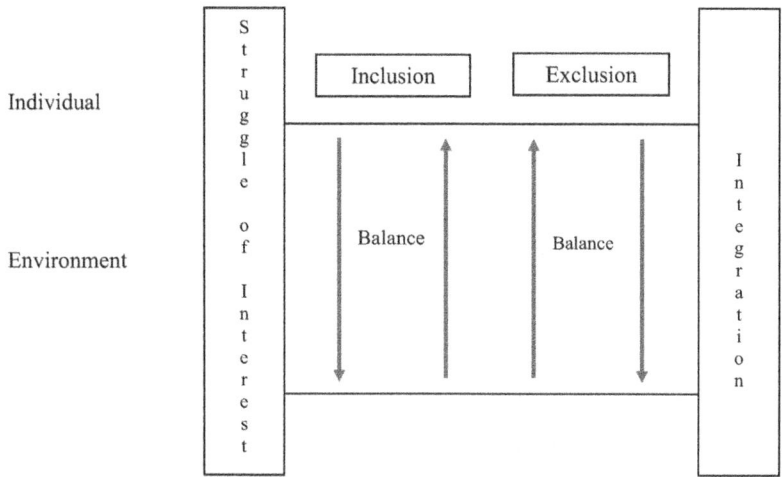

Fig. 1.1 Internal relations between the struggle for interests and integration

*The involvement/disengagement in the historically constantly changing so-
cial struggle of interests or balance of interests on the way to social integration, fed by the anthropologically determined situation of man, who is forced
in the way of acting, thinking and language to constructively develop his
forms of life himself.*

Since action, as has already been argued, is self-determined, the normative
specification for participation in life, at least in contemporary democratically constituted societies, can be understood as a self-determined life filled with meaning.
The analytical distinction of participation in different societal spheres will be discussed in detail below (Fig. 1.2).

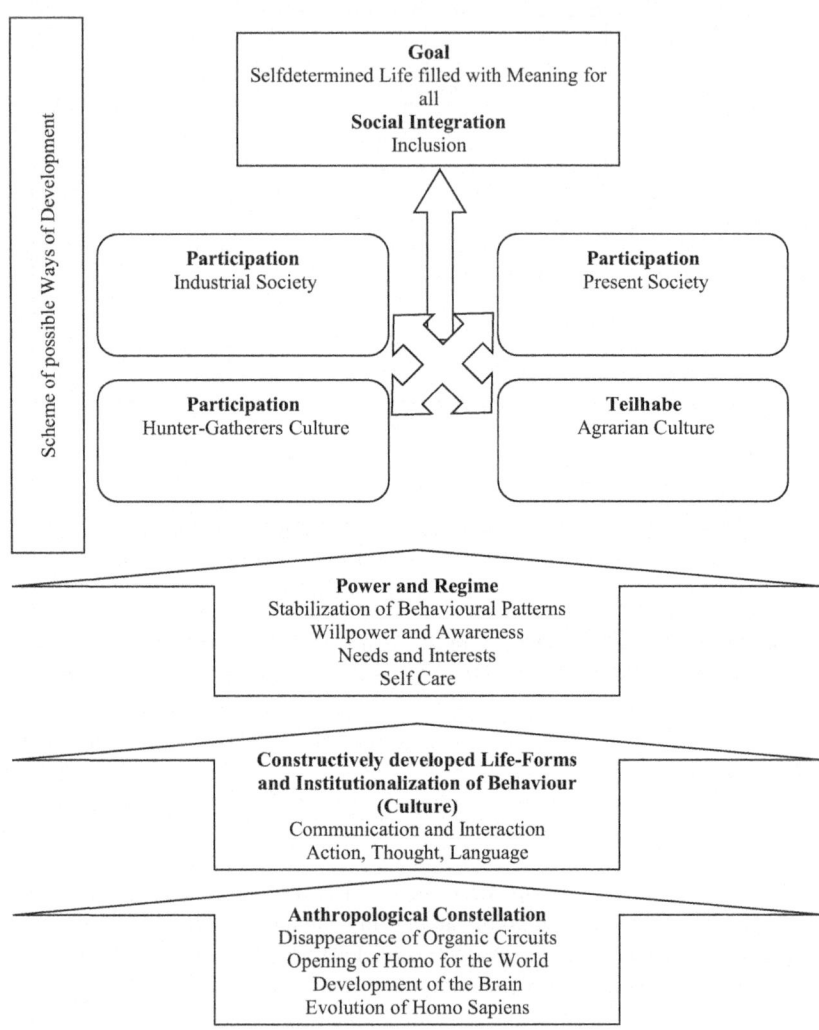

Fig. 1.2 Forms of life starting from nature

General Concept of Participation

2

For the further discussion, I have proposed to understand participation as follows: The involvement/disengagement in the historically ever-changing societal struggle of interests or balancing of interests on the way to societal integration throughout the life course, fed by the anthropologically determined situation of the human being, who is forced in the way via action, thought and language to constructively develop his forms of life himself. Since action is self-determined, the normative specification for participation in life, at least in contemporary democratically constituted societies, can be understood as a life that is self-determined and filled with meaning. This proposal now needs to be expanded with a sociologically familiar repertoire of terms.

Participation is action that starts from resources, is dependent on social forms of organization in its execution, and produces its effect in the context of social integration. I am explicitly not concerned here with the individual act of action, which would neither be comprehensible in everyday understanding nor fruitful for scientific observation. Rather, the focus is on actions that follow on from one another, have become normatively stabilized over time, and have thus become forms of behavior. Action is controlled by potentials or resources that can be found both in the individual and in the environment. Such action takes place in social settings (forms of organisation) in which they generate meaning and, through their repetitiveness, in turn contribute to the consolidation of social forms of organisation.

Integration into society is a universal requirement that virtually characterizes the constitution of society. The goal of social integration is an independent life filled with meaning, at least ideally, in the context of Western democracies.

Conceptually, therefore, participation is to be outlined in a first step together with social integration, since only participation leads to social integration, the latter being the result of the former (integration that has already taken place can strengthen participation). Secondly, it stands in a framework of standards of living (the term was used, for the first time, by Otto Neurath translating the german "Lebenslage"), which contain the resources and also offer the social forms of organisation, clearly visible, for example, in the voluntary activity in clubs, associations and loose groups. Standards of living are constellations of external conditions (external resources) that people encounter in the course of their lives; they are closely related to the cognitive and emotional patterns of interpretation and processing (internal resources) that develop in interdependence and which people produce. Standard of living is a dynamic concept that encompasses the historical, social and cultural change-generating development of external conditions, but also the specific forms of interaction and possible areas of tension between people's actions and these external conditions. Moreover, it should be noted: Since standards of living always involve the distribution of unequal social situations, the conceptual framework of participation also includes the dimension of social inequality. According to all empirical findings, only sub-groups of older people have good preconditions for the realisation of participation, so the inequality dimension is factually indispensable. Thirdly, the political dimension must be taken into account. Ageing populations represent one of the great political tasks of the present day. Institutionally and legally, this task lies primarily in the area of social policy. From this point of view, there is a strong link to the issue of participation in so far as the activities of social policy are concerned with the shaping of living conditions and, in this context, create or should create the appropriate framework conditions for participation in old age, whereby the change in the conditions of social inequality can be regarded as one of several objectives for action. Fourthly, it must be taken into account that participation is an expression of an indissoluble individual-environment relationship, so that individual (internal) and structural (external) resources must always be assumed. This condition can again be linked back to the standard of living, because it is in these that the margins or scopes of action are given and the areas of tension within which the opportunities for participation are distributed become visible.[1]

In the concept developed here, the focus is on participation; its objective is social integration, or in the case of enhancement, social inclusion; its preconditions are to be sought in the always unequally distributed standards of living. They offer scope, which contains the resources and is decisive for the different forms of participation. For their part, standards of living are essentially shaped by social policy. Any change in participation conditions is inevitably shaped by the individual-environment relationship, which in the context under discussion here implies the idea of the everyday world. Since participation, integration, living conditions (including scope and resources) as well as socio-political design and individual-environment represent the main building blocks of the concept, it is conceptually necessary to line these in turn with a terminology that is as clear as possible.

2.1   Participation and Integration

In terms of their meaning, participation has to be thought of as a general but manifold notion; on the one hand, it means being part of something larger, a community (inclusion), on the other hand, it contains the idea of possessing or wanting to get something. "Ultimately, [in participation] there is also an emancipative claim that aims at improving social conditions as a whole" (Heimgartner and Findenig 2017, p. 21). To understand participation in its full sense it must not be seen only as *social* participation, as it is often the case (despite theoretically differentiated gradual stages). In that narrow understanding it is underdetermined, mostly unclear and consequently, as a so-called "container word", also remains empty in terms of content, thereby rather insinuating an abstract form of participation, in which it seems

---

[1] Here it is useful to add a supplement. Although psychoanalysis has repeatedly been accused of absolutizing the organism or the individual in a one-sided view (Michael Balint said that the formation of concepts in psychoanalysis, apart from the exceptions of the terms "object" and "object relationship", only refers to the individual), its offers for the topic of the individual-environment should not be underestimated or even wiped off the table. The establishment of psychic structures, even in Sigmund Freud's classical understanding, is inconceivable without the relational context of the subject-object par excellence; terms such as superego, reality testing, introjection, object relation, etc., could not be meaningfully integrated into theory without the "external reference" of their meaning; some authors, such as Yvonne Schütze (1982, p. 129), have used the transformations of psychoanalysis to virtually underscore the importance of the individual-environment interaction in the development of the oral phase.

difficult, for example, to assign an appropriate place to the sensation or feeling. Participation often appears as worn down like a proverb used by generations.

In empirical research on ageing, a distinction is traditionally made between *economic, political, cultural and social participation* (Kuhlmann et al. 2016). Critically, it should be noted that in the discourse on social gerontology and senior citizens' policy, the concept of participation is not uniform and is usually defined very narrowly, with the focus often being on civic or voluntary engagement. However, it is also true that valid empirical results can be achieved with a typification such as the one mentioned above, with appropriate subdivision of the individual areas into dimensions and their measurement by indicators. For example, findings that have been stable for some time include that those older people engage more in participation who have already done so in their lives to date and have been able to establish the necessary action logics[2] (Aner 2005; Erlinghagen 2008), or that older people with lower incomes and lower formal educational qualifications are less likely to engage (Aner and Köster 2016, p. 476). Nevertheless, it seems urgent to start with a systematic compilation of such stable correlations, to reformulate the individual findings into more general hypotheses in order to arrive at richer results on a meta-level on which the newly generated hypotheses could be tested again (cf. as an attempt: Sect. 5.3). A procedure that was once already proposed by Otto Neurath with the intention of condensing empirical knowledge (Neurath 1981). Today, this idea is pursued in meta-analyses, which are mainly conducted statistically, but is very limited by this logic, which focuses on statistical procedures, and by the types of data required for them. An alternative would be offered in the context of first- and second-order sciences, which, however, is much too little noticed (cf. Malnar and Müller 2015).

In recent years, there has been an increasing number of projects in which participation in old age has focused on certain groups who, according to my conception, are in difficult or clearly disadvantaged standards of living. These included, for example, people in poverty, especially poor women, people with disabilities, with a migration background, with care needs, with dementia, etc. In these projects it becomes apparent that a concept of participation tailored to everyday normal life situations cannot do justice to all important aspects of the lives of the disadvantaged. "Normal" is understood here in the sense of Alfred Schütz, in which the everyday living environment is experienced as simply given and as unproblematic until further notice. If the unquestionable is called into question, this requires its

---

[2] The term logic of action appears frequently in studies, but remains strangely underdetermined. It would certainly be helpful if it were incorporated into an elaborate theory of action, such as that outlined by G. Dux (Dux 1992).

own treatment. Understandably, perspectives and conceptions of social work and social pedagogy come to the fore in these projects, which, however, are sometimes somewhat distant from established sociological-gerontological hypotheses. The fact that this is also linked to different traditions of understanding science and practice, in which divergent horizons of understanding problems become effective, has been shown on various occasions (Amann et al. 2010b). Here, the task of developing appropriate research programmes arises in a special way.

In the social sciences, it is advisable not to look for measures of perfect integration of people into their society, because such a notion will fail on the measure of perfection. There will never be perfect integration for all, because there can be no society without inequality. The problem in this whole discussion, after all, has always been not the desire for a better world, but the belief in the utopia of a perfect world (Hobsbawm 2004, p. 215). It is therefore useful to choose a target for social integration of which it is clear from the outset that there can only ever be an approximation, which can only be determined as more or less (to say that someone is twice or three times as well integrated as someone else is scientifically meaningless). I want to see this target in the possibility,

> *to lead a self-determined and meaningful life, where meaning is derived from access to and effective use of all available social conditions that can be conducive to integration.*

Obviously, it is about questions of access opportunities and their feasibility, as they were already discussed by Ralf Dahrendorf in the 1970s (Dahrendorf 1975). Here, the concept of social integration is combined with the concept of standard of living and their respective specific margins or scopes. In the capitalist market economy, the integration of people into their society can only be achieved through the labour market, because this is the only way to achieve the means that enable and define active participation in social life. This can be broken down into the simple basic figure that some people have to sell their labour power at the respective market price, while others have to make the greatest possible profit with their property or possessions. The fact that there are millions who can nevertheless participate in some way without work or property has to do with the fact that in those cases where someone does not belong to one of the two groups, society (the welfare state or other institutions) steps in. The areas in which the above-mentioned preconditions are met, in which participation of a material, social, cultural and political kind becomes possible, are manifold, because the system that appears as a market economy and corresponds to the democratically constituted state is highly differentiated, it tends to open up more and more possibilities and forms of participation for individual lifestyles, albeit under special conditions in old age, and thus also for integration. And yet these options are by no means accessible to all (Amann 2017).

2.2   Standard of Living

The short definition of standard of living (Amann 1983, p. 147), which is meanwhile cited very frequently in the relevant literature, already contains the essential provisions in its core. Standards of Living are constellations of external conditions (external resources) that people encounter in the course of their lives; they are closely related to the cognitive and emotional patterns of interpretation and processing (internal resources) that people develop in mutual dependence. Standard of Living is thus a dynamic concept that encompasses the historical, social and cultural change-generating development of external conditions, but also the specific forms of interaction and possible areas of tension between people's actions and these external conditions. In gross simplification and without explanatory remarks, the following should be noted. The external conditions of life are the economic, social and cultural conditions, which in turn are created by the mode of production, the division of labour, the differentiation of occupations, the institutions of social and political power, and the distribution of privileges. We can without difficulty identify in these external circumstances all the matter that determines our lives: e.g. from the conditions of employment and income to legal regulations of the state to moral requirements of religious communities, and from the social forms of organization of networks to formal organizations such as clubs, associations, and so on. In order to do justice to the two time dynamics that both Otto Neurath and Gerhard Weisser had underlined, I have emphasized two perspectives: on the one hand, those of successive cohorts, each of which finds different heights in the distribution of social wealth and social and political freedoms (or unfreedoms) under which they then have to live, and the actual and potential opportunities for access and disposal that result from this; on the other hand, the life-course-dependent scope that the individual finds and actually exploits in each case for shaping his or her existence. Finally, it is to be assumed that although standards of living affect individuals or groups in their totality, the inner configuration, conversely, can neither be perceived and interpreted by the individual in its totality, nor can it be examined without separation into individual dimensions.[3] Therefore, the concept of standard of living *dimensions* suggests itself.

In order to characterise the inner principles of the design of leeway in more detail, it is useful to start from the premise of scarcity and balance-work, constraint and autonomy. Scarcity is a constant relation between objective availability and subjective ability to dispose of (almost) any given thing, whether time, money or forces. It forces coordination, which means "work", because no one has unlimited perceptual and processing potentials in the mental-psychic realm, nor unlimited economic means and physical forces at their disposal (Amann 1990, p. 181). If this idea is narrowed down to exemplarily selected important areas of interaction, three can be found: The professional and working world with production, work organization and performance control; family as a quasi-opposite world to the former; finally, the "free" social relations with their probably lowest degree of institutionalization of normative requirements. In these fields, people must constantly calculate, distribute and deploy energies and forces. In some, coercion is more in the foreground, in others autonomy; people have to do *balance work* (Amann 1990, p. 181). In order to be capable of the necessary separation or combination of temporal, factual and social demands, people must be socialised accordingly. Thus, in the course of life, varying more or less well designed action requirements and action competencies develop, which are the basis of balance work. "Requirements, expectations, offers, and dispositions are each specifically structured and highly prioritized within the three fields mentioned" (Amann 1990, p. 183). Choices and decisions are never made completely arbitrarily; they are tied to socially predetermined *priorities*. Individuals' modes of perception and action are inextricably linked to these priorities, and the totality of all these factual, temporal, and social circumstances define *scopes of action*. Against this structure of unequally distributed priorities, which are generally of a stable nature, we have to assume subjective scopes of disposition which make it possible to act according to one's own judgment. "We can therefore assume with certainty, on the basis of the necessity of a constant balancing work between the three areas, that this balancing work is organized by man in a structure of dispositional scopes which he [or she] learns to use throughout life (...) Linking external conditions and inner autonomy, we can thus speak of *learned dispositional scopes*" (Amann 1990, p. 183).

At this point, the idea addressed above with the concept of lifeworld is to be expanded. Up to now, research on living situations has mainly been designed as situation research, i.e. empirical-analytical with a focus on external conditions,

---

[3] Which is not to say that people do not reflect on their life situations. Here we would have to deal with the problem of a first and second order observation.

supplemented by subjective assessments and evaluations of these external conditions. A conception of "small life worlds" (the term goes back to Benita Luckmann), however, would now have to be thought in terms of communicative life worlds. These lifeworlds can be ascribed a generative structure, since they are based on a special type of experience, action and knowledge (Soeffner 1989, p. 15), and they are conceived as spaces of interaction due to an original competence of the actors. These spaces of interaction are to be grasped on the basis of "excellent" facts, which are found above all in the reduction to communicative action. These facts are: (a) the social relations in which the "other" has a primacy (in the everyday world) for the constitution of meaning; (b) the social consciousness (knowledge), which is stratified according to horizons of meaning and contains intentionality; (c) the social meaning, to which, in a strict understanding, only the phenomenological reduction leads back, which is then able to open up the diversity of everyday understanding of meaning; (d) the communication, which is reflected in the typology of life worlds. It goes without saying that an integration of the concept of standard of living, as discussed so far, with a concept of communicative life world would first require the determination of the epistemological and scientific-theoretical preconditions as well as a rigorous conceptual analysis – which cannot be done here.

The core of the standard of living concept can therefore be described as a dialectical relationship between circumstances and behaviour, whereby this relationship simultaneously has a conditioned and structured as well as a conditioning and structuring side. "Standards of living are dynamic in the perspective of their permanent social, economic and cultural change; they are persistent in the perspective of their conditions, which can only be changed through effort" (Amann 2000, p. 58). All standards of living are an expression of socially produced systems of inequality, "in them the mediation between structure and the subjects who productively process their reality becomes manifest" (Amann 2000, p. 58).

Ingeborg Nahnsen was the first to attempt to flesh out the idea of scope of action in this context, describing it as a "total concept of the social opportunities of the individual" (Nahnsen 1975, p. 148). This was not a simple shift in terminology, for it was necessary to delineate the framework within which social-structurally relevant dimensions and the hypotheses that might link them could be determined. This framework encompasses a multitude of conditions to be thought of analytically, which have to be arranged according to complexes "or, in other words, combined into several (fictitious) individual scopes of standard of living" (Nahnsen 1975, p. 150). The actual selection from the large number of possible fictitious scopes then leads to the following, in which the degree of possible development of interests and realization of interests depend on the scope of action specific external conditions:

1. *"Supply and income scope* (extent of possible supply of goods and services).
2. *Scope for contact and cooperation* (opportunities for maintaining social contacts and interacting with others)
3. *Scope for learning and experience* (conditions of socialisation, form and content of the internationalisation (sic!) of social norms, educational and training fate, experiences in the world of work, degree of possible occupational and spatial mobility)
4. *Scope for leisure and regeneration* (psycho-physical stress due to working conditions, living environment, environment, existential insecurity)
5. *Dispositional scope* (circumstances on which it depends how significantly the individual can participate in decision-making in the various spheres of life)."

It is obvious that this categorisation of scope has been subject to many additions, criticisms and attempts at empirical implementation to date. It is also obvious, however, that in some subject areas it coincides quite clearly with those that are considered relevant for participation research.

The development so far has led to the fact that concepts of standard of living have come into competition with other concepts, such as those of social structure analysis, the analysis of social situations, and also that of quality of life; in some cases they simply overlap. It is evident that this process was partly determined by the abandonment of older models of stratification and class. That the term standard of living is "by no means unambiguous and selective" (Schwenk 1999) can best be accepted from this perspective. However, I do not want to go further into this extended lamentation here, because all too often it is precisely the harsh criticism that lacks the expected constructive continuations. Rather, the following requirements should be emphasised: (a) to gain a systematic overview of the empirical diversity of dimensions in standards of living research, and (b) on this basis to attempt to increase the complexity of the concept. Standard of living concepts that are to be valid as such must at least

• cover different structural levels
• preserve the trans-economic diversity of the aspects of reality
• consider the time dynamics
• take into account the participation or design aspects or potentials of the individual, and
• allow judgements to be made about more or less pleasing or advantageous situations.

In the analysis of standard of living, the actual situation of each individual person is usually considered; this applies in any case without restriction to all studies in which survey data are used, whether they are primarily collected or are used for secondary analysis from existing data sets. It is equally clear that some dimensions and the indicators defined for them (the problem of non-direct observability) are used with constant regularity (such as income, education, gainful employment, health, etc.), while others are used very rarely and in a rather colourful selection. Nutritional status, available clothing, personal hygiene facilities, etc., for example, are likely to be used only in research on living conditions and poverty. This is probably related to a phenomenon in the sociology of science, which has to do with the relations between theory or conceptualization and data. In any survey, starting from the conceptualization to be theoretically argued (which dimensions are chosen), a selection from reality takes place, which in turn is co-determined by the anticipatory view of the desired and achievable data. This can go so far as to exclude from the conceptualization questions that experience has shown to be tricky (that produce poor data validity or give rise to fears of high non-response rates). The empirical result then represents a section of reality within a certain time segment.[4]

2.3   Social Policy/Senior Policy: Reconsideration?

I will start with a comprehensive view. In Austria, the term "*social security*" covers a wide variety of regulatory matters. Here, those areas are to be considered that have a specific relation to the older population. Since, apart from the Social Report of the Federal Ministry of Labour, Social Affairs and Consumer Protection (BMASK), there are hardly any relevant social science studies, this report provides the basis for the systematics. In a rough classification, the following are to be mentioned:

- Social security (especially pension insurance)
- Unemployment insurance, labour market service
- Universal systems (care provision, de facto effect)
- Means-tested benefits (compensatory allowances and minimum benefits)
- Social protection for civil servants
- Social compensation
- Employment law safeguards

---

[4]These spatio-temporal limitations are clearly evident in the presentation of the dimensions and indicators with which participation is empirically measured.

- Occupational forms of pension provision
- Social Services.

In a sociological understanding of the topic of social security and social protection, *social policy* is of particular importance. It is an instrument, and therein lies the significance of *social protection,* with the help of which intervention is made in the standards of living and living conditions of individuals or groups of people in order to reduce risk or threat situations or to prevent their occurrence. Here I follow the views of Otto Neurath and Gerhard Weisser (cf. Amann 1983). Indirectly, it thus also influences the (external) resources from which participation emerges. Since the conception of standard of living used here is an expression of the social situation, there is a direct line to the tasks of social policy as a shaping medium also from the perspective of social inequality. Its measures should serve social cohesion and the management of social, demographic and economic change. This also includes the prevention and avoidance of poverty. In Chap. 8 I will deal with strategies for the prevention of poverty in old age by way of example. Under this objective, senior citizens' policy must also be generational policy.

Inevitably, the question arises here as to what politics/social policy should and could achieve under the given circumstances. It is not uncommon to get the impression that people even expect politics to establish solidarity between the generations, the integration of the older cohorts and the like. However, it would be a gross misunderstanding if anyone expected politics to be able to produce such attitudes and modes of action in people and to enforce corresponding attitudes. Politics can only refer to external framework conditions. Developing perspectives for a future policy for senior citizens requires, *first and foremost, a realistic assessment of what politics can be and what its task would be.*

Instead of summarising numerous analyses from philosophical and political science compendia on the nature and task of politics, I would like to make a pragmatic proposal here as to how these questions could be answered and how a specific understanding of senior citizens' politics could be argued. *Politics is the shaping of social conditions by democratic means;* what politics understood in this way aims at are framework conditions, external circumstances of people's social existence, i.e. standards of living. The means it uses are those provided for in a democratic community. In this understanding, social policy for older people or senior policy would then be the shaping of the living conditions of ageing and old people. It should be noted at this point: The means currently used by politics for the older are not all that could be exhausted within the framework of our democratic systems.

After a preliminary definition of what senior citizens' policy could be and what its task consists of, we must now ask about its scope. A first difficulty for the determination of the scope in terms of its subject area arises on closer inspection from the fact that at present there is no separate policy for senior citizens. The securing of pensions is, at least legally and institutionally, a question of pension insurance policy; care provision in old age is a question of the design of the Federal Care Allowance Act, home laws, etc.; the practice of helping and caring for older people in almost 80% of all cases is a matter for families or, unfortunately, tends to be unequally distributed still always only for wives and daughters; the situation of older workers is a question of labour market and employment policy and, in turn, also a question of social security. The approach to new perspectives on senior citizens' policy should therefore be chosen along a path that does not adhere too closely to the institutionalised distribution of tasks and competences. Instead, it would be necessary to move towards

> *seeing the tasks of a comprehensive senior citizens' policy on the one hand from the changing living conditions of the ageing people themselves and on the other hand from the intergenerational development.*

Only this double view can show us that for a prosperous development of the living conditions of the old, the safeguarding of pensions must be just as much a concern as a successful employment and youth policy (the epitome of generational policy). Young people who cannot find an apprenticeship, who cannot get a job after further education, who are unemployed and without prospects, will hardly look back on a fulfilled life as seniors and feel solidarity towards a society that has spoiled their "best years". At this point, an often little-noticed fact needs to be emphasized: The policies made yesterday by today's senior citzens determine our lives just as much as those we make today will determine the lives of tomorrow's old.

What must unfortunately be noted is the fact that demographic change and its consequences in Austrian politics, with the exception of the pension and long-term care discussions, have not yet led to any comprehensive discussions and programme drafts appropriate to the situation. In fact, and this can be known, the situation presents itself in very diverse ways and completely different from that of 30 years ago. The demographic development can be outlined, for example, by the following trends: there are

• a so-called collective and relative ageing of the population, as well as a considerable and rapid increase in the number of very old people

- the structural changes in the ageing phase (including individualisation, feminisation and pluralisation of old age)
- a population which would be shrinking very strongly without immigration
- Heterogeneity and internationalization of old age with increasing ethnic singularization and alienation
- Segregation and inequality in old age
- Island formations in contraction and growth (economy) as well as in employment opportunities (cf. Hüther and Naegele 2013).

The most important problem, however, is that in Austria, although social policy falls within the agendas of the federal government, the provinces and the municipalities, a corresponding differentiation in the public discussion or even the relevant research is virtually non-existent. Austrian municipalities and regions are not only ageing in terms of population average, they are often shrinking, becoming more heterogeneous in people's living conditions, subject to processes of singularisation and segregation.

In para. 7.2.2 I will deal with the concept of the life course in more detail; this is where an anticipation of family policy has its place. In view of the current political discussions in Austria on family, education, upbringing, etc. (including the so-called "daddy month"), it would be worth considering whether it might not help to rethink the family, the objectives of which could be found in a rethinking of the relationship between social change and the future viability of the family, in which the life course would obviously offer a profitable perspective. It suggests a longitudinal view and refers to family as a process and as three-dimensional in its modalities: it must be seen as an institution with changing personal references, as a segment of everyday life with changing external ties and linkages, and it must ultimately be seen as a kind of barometer with phase-dependent surges of modernization or traditionalization (cf. Krüger 2010). The life course perspective of gender draws attention to different opportunities for participation in the labour market and family, in which opportunity structures shift continuously between occupations, labour market segments and educational opportunities, both individually and across cohorts. The family is also constantly interlinked with life-course-relevant institutions, the most common of which is the education system, which helps to determine the life course of the offspring in each case up to retirement age. But it also has to deal with kindergartens, hospitals, homes for the old and nursing homes, which in turn have to be regarded as institutions of social policy. They all enter into relationships with family members that are dependent on life stage and age. Many other institutions, from doctors' surgeries to shopping places to offices, government agencies, and transportation infrastructure, shape the relationship between family (privacy) and public and contribute to family life differently at different family stages (Krüger 2010, p. 229). The cardinal question, however, concerning all these relations and relationships is: to what extent are they all actually attuned to the manifold concerns of family life, to what extent do they negate them, in that institutional and party-political self-interests are stronger?

However, the scope of senior citizens' policy is not only determined by its subject matter, i.e. the changing living conditions of ageing people and intergenerational development, it also relates to the democratic means I have already mentioned. Here, a new policy would have to involve senior citizens themselves in opinion-forming and decision-making processes to a much greater extent than has been the case up to now, and grant them a seat and a voice in the most diverse groups and representations. Last but not least, the role of the senior citizens' associations and federations, their offers and their contribution to the activation of the older people as well as their structural resources for the commitment to participation would have to be considered more closely here. Of course, this argument also extends to the issues of a new culture of ageing. Politically active seniors who help shape their own affairs do not appear out of nowhere after retirement; they must have already been interested and active in their previous lives, they must have already experienced and learned the relevant requirements and opportunities – and this from childhood and youth. But this presupposes a population that *does* not rely solely on what others do for them – politics and *the* state. Seen in this light, a new senior citizens' policy is a long-term project that must be embedded in a change of political culture in this country, in a different awareness and other possibilities of citizenship.

In the conventional understanding, policy for older people, far ahead of all other objectives, is social policy in the classical sense: under the heading of "social security" all measures for the case of "old age" are meant by it. This conception dates in its foundations from the time of the developing welfare state, which began to establish itself not in a "greying" but in a relatively young population or society. However, ageing societies, such as the industrial ones at present, also present the state and politics with other tasks.

In the opinion of many people, a rethinking of the intergenerational contract far beyond the monetary aspects and the safeguarding of the material basis of life, which must remain an integrating core element of a senior citizens' policy, is part of it. That this "contract" needs some corrections is no question, but the opinion of those who want to question and abolish it at all only shows a fundamental ignorance of the functioning and stability requirements of the democratic system within the framework of a capitalist market economy. It must also be about a fair distribution of burdens and duties, about active cooperation in the multitude of growing generations, about new forms of encounter and common fields of action. Forms of communication and models of cooperation must be developed, civic commitment must be awakened and promoted, family systems must be supported and their services supplemented (e.g. in the expansion of childcare facilities and cross-state minimum standards in child and youth welfare). In this context, should not the entire system of transfer payments, for example, be examined and critically reviewed in order to be able to assess whether the overall effect achieves the promotion of family systems that was intended and hoped for? Would it not make sense to switch from a policy of patronage to a more targeted policy? There is an urgent need to set a decisive course for the future (cf. Chap. 8). To repeat: a policy for senior citizens will neither be able to decree this change nor to conjure it up, but it must create favourable and encouraging conditions for it.

The broad field of health, care and nursing could also become the subject of a senior citizens' policy to a far greater extent than has been the case to date. In many respects, additional insights and knowledge are not directly necessary to prove how important prevention and geriatric-gerontological rehabilitation are in order to improve the quality of life in old age, to reduce costs and to meet the requirements of an ageing society in a specific way. For example, would it not be the task of a new senior policy to make the topic of very old age and its consequences one of the main topics – a number of important aspects would be put into a different light here, starting with isolation and loneliness, to the need for care in the case of dementia and fragility, financial disadvantage, precarious housing conditions and fundamental disadvantages, to the predominant share of women in the old age groups and their special worse situation in life? Last but not least, overcoming the sometimes very obstructive distinction between social and health care systems would also be useful.

2.4 Individual-Environment Relationship

At this point, there is no longer any need for a separate justification of the argument that the individual and his or her social world are indissolubly linked; this has long been done anthropologically and in terms of action theory. However, the question of how this relationship could be successfully conceptualized in the context of participation in old age is far from being answered. Proposals range from "person-environment-fit" as attempted by Robert D. Caplan (Caplan 1987) to more recent models related to residential and technological environments (Oswald and Wahl 2016). The proposal made here, as already stated, is closely related to the standard of living concept and there specifically to the notions of "learned dispositional scopes" and "resources", although it should be additionally emphasized that I am speaking here of a specific person-environment relationship that holds alternate correspondences (or their opposite) that do not apply to all populations. Moreover, the focus is on the relationship rather than the individual or the environment.

In a wider context, these considerations have been confirmed in an EU project carried out in eight European countries and coordinated in Vienna by Anton Amann and Ralf Risser, whose main theme was mobility and quality of life in old age. The "wider context" was established by relating individual needs and desires to environmental conditions of a physical and social nature. This combination then led to "profiles" of what older people want or demand, which could be named as follows:

- *Safety:* It is understood as the avoidance of situations or events that can lead to harm to the individual (here the distinction between safety and security is significant).
- *Accessibility:* It is the state of full accessibility and usability of public space, services, aids and movement.
- *Comfort:* This refers to a combination of well-being with the conditions of action and situation within immediate reach (lifeworld) that are relevant to sensory experience.
- *Attractiveness:* it means the extent and quality of attraction by the public space in relation to the ability to start and sustain activities and to achieve social involvement.
- *Intermodality:* it is the integration and interchangeability of different ways and means of activating resources and using them in action (the idea of compensating for limitations)
- *Technological Fit:* This means, in accordance with the most modern developments, the adaptation of the material-technical environment according to standards of a new and successful development of real time information (Internet, SMS etc.) (SIZE 2006).

From a sociological and equally from a psychological perspective, these "profiles" are specific results about knowledge and relationships of older people. They require contributions from the environment that are oriented towards the needs of the individuals, but they obviously also require that older people be included in the construction processes, that the consensus on measures also be extended to them. They themselves can contribute to answering the significant question in terms of planning: What specific preconditions does a concrete network of knowledge and relationships offer?

The environment should be divided, in the Schützian sense, into directly accessible, indirectly accessible and distant, more difficult to reach environments, whereby it would be obvious to call the first level the social co-world. The environments include people, objects and information. The individual can be conceived as that unit in which perception, experience, emotional attachment and mental coping with the environment take place by way of processing-learning and applying: social education. Social education also includes the understanding that people do not exist in a vacuum and are continously influenced by a constant interaction of both (or more) sides (Sting 2010). The constant interaction between the two sides can be conceived as that relationship through which identity, autonomy, well-being, satisfaction, etc. can be endowed on the part of the individual, and design, change, transformation on the part of the environment. The following figure may schematically illustrate the relationships formulated here, whereby the insight is self-evident that the scheme includes preconditions for participation, participation as a specific activity, and consequences of participation.

The complexity of all the interrelationships shown in Fig. 2.1 makes it impossible to attempt to empirically represent them in a single research project. The need to develop appropriate models for this with the aid of time series analyses, multilevel structural equation models and hermeneutic approaches, etc., points to the need to develop an elaborate research programme with a long-term perspective.

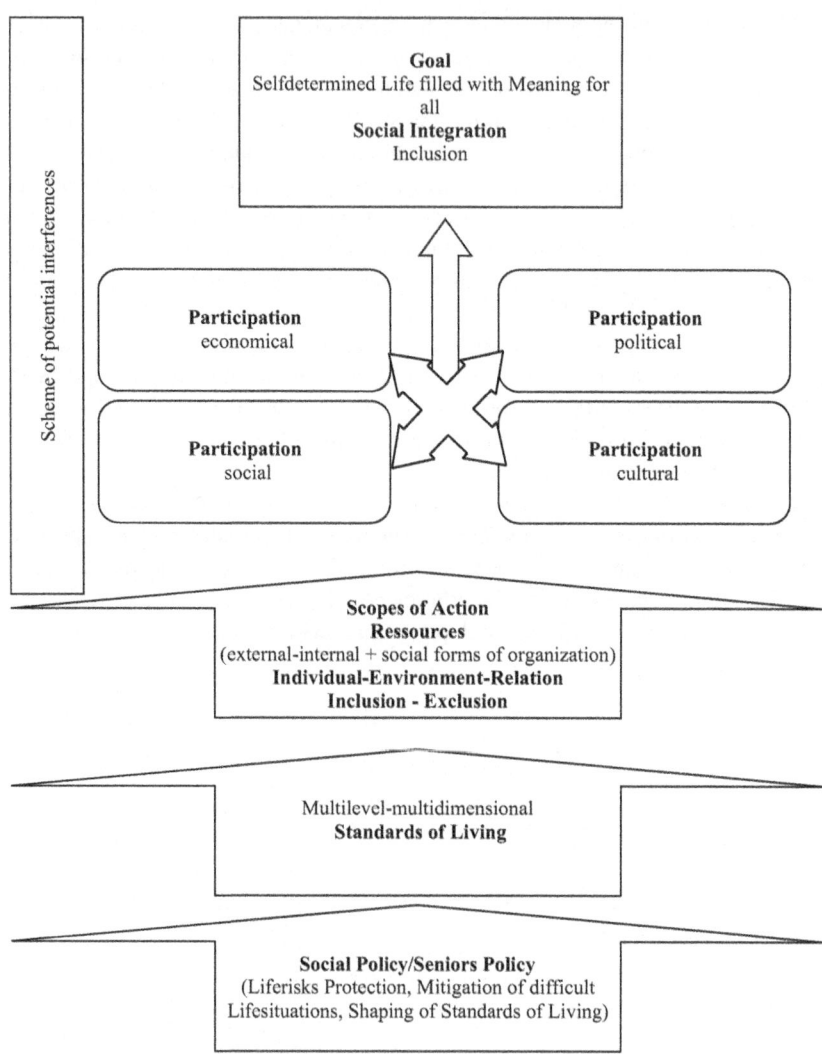

Fig. 2.1 Interdependent structure of participation in old age

Part II

Social-Gerontological Centering

Participation in Growing Old

Before moving on to the narrower issue of participation in life in old age, a general sociological question remains to be discussed. In the attempt made above to reconstruct the possible course of development in which human beings constructively produce and had to produce their own forms of life, the sociological core lies in the question of how the interplay between subject and environment is constituted, in which each individual human being in his ontogenesis can acquire and form the competencies of action, language and thought in such a way as society makes possible in each case. This touches on the time-honoured dichotomy of individual and society, which from earliest times – e.g. with Protagoras of Abdera – up to the present day has not been able to find a satisfactory resolution in sociology. For the time being, the question of the relationship between man and his society should mean no more than that two conditions of sociological possibilities of knowledge should be considered more closely: The moments to which sociological theory must link up if it wants to understand man in a society which he and which at the same time produces him, and the conditions under which human cognition in the sociological sense becomes possible in the first place. As has already been shown, this involves at least questions of so-called social theories, but also questions of an anthropological-sociological justifiability of the subject and of cognition (of which theory is, after all, a part). Since the beginning of the modern era, attempts had been made to lay the foundations of philosophy and the sciences in an epistemology that was supposed to contain the methodological principles of secure knowledge before any determination of the content of knowledge. As the nineteenth century drew to a close, however, it became increasingly clear that such knowledge could not arise from itself, beginning at a final starting point that had been secured once and for all and then ascending step by step to a formulated theory of knowledge. It became increasingly clear that all cognition is involved in a comprehensive context

A. Amann, *Living - Participating - Growing Old*, https://doi.org/10.1007/978-3-658-39681-7_3

of life and arises and changes in the context of practical human activity. Later, Jean Piaget was able to say that experience is mediated by action. This insight required a deepening of the epistemological foundations, which were now sought in a philosophical anthropology. Corresponding to this development was a second one, which, also beginning in the nineteenth century, became even more momentous for sociology. The understanding of the human life-world was put into terms and categories which have retained their binding force to this day: It must be understood, first, as the result of the actions of socialized subjects, and it must be understood, second, as the specific social formation to which these subjects always already belong. In this way, social theory was almost unmistakably anchored in action theory (and thus, of course, a theory of the subject) which was placed in the (coercive) context of society. It is precisely for this reason that it is understandable that thinkers such as Karl Marx, Émile Durkheim, and Max Weber attempted to conceive of a theory of society as, so to speak, a generic history of man. As incomplete as all previous attempts have remained, the program stands and awaits its redemption.

3.1   The Malaise of a Time-Honoured Dichotomy

As insightful as the fact that it was only with sociology as an empirical science that this question of the fractured unity of man and society and its discussion were shifted to the level of empirically observable individual and collective action, so too is the fact that sociology has repeatedly failed to answer it. The separation between subject and object as an act of reflexive distance continues in the separation between practical action and theoretical reflection (it is also from this separation that Immanuel Kant's dictum can be understood that nothing is as practical as a good theory), and again finds its extension in the separation of philosophy and science, and finally in the distinction between individual special fields in the sciences – in sociology: "general" sociology as the field of grand theories, "special" or "applied" sociology as the fields of its problem-oriented empirical research, and finally "methods" of sociology as the field of scientific tools. This division has above all a practical sense, which is in line with the historical emergence of this science: on the one hand there are the theoretical questions, the fundamental questions of sociology, on the other hand the substantive individual moments (hyphenated sociologies), connected with practical skills, as they have to be appropriated in the sense of social usability. This distinction, however questionable, also expresses the dual character of sociology: As theoretical sociology, it was always on the knife's edge of also being philosophy of society, philosophical ethics; as empirical-practical sociology, it is that complex which developed out of the cameral sciences, especially out of eighteenth century mercantilism, out of the needs of the emerging central states, when

for the first time something like points of view of a planned economy and administration emerged, for which then also overviews of all possible needs, structural relations and desires became necessary. Sociology is thus – quite in contrast to the relatively closed doctrinal buildings of many other disciplines – an inherently
contradictory science, which on the one hand is committed to the immediately socially useful in moments of detail and craftsmanship, while in its theoretical-reflex-
ive claim it is at the same time directed to the conditions lying behind this immediate
visibility of the social – to "what holds the world together in its innermost core". It
is precisely this last claim, however, that easily leads to a concept of theory that
grasps theory as something abstract in relation to the individual social moments, and
thus itself becomes problematic. As a rough thesis, let us note that the contradictions
in society that determine its dynamics are reflected in sociology as a social and at the
same time cognitive system, and thus sociological theory itself must be understood
as an expression of social conflicts. This particular situation has a specific consequence for the way in which sociological knowledge can be mediated or imparted;
after all, it is necessary to keep in mind the enterprise of understanding what is to be
mediated, as well as the process of mediation itself, still as a moment of a social
process to which the theory belongs. In all beginnings, says Ernst Bloch, there is a
character of arbitrariness; the principle of constantly inquiring curiosity, as contained in the image of the "Faustian" man, does not dispense with this.

Theoretical thinking is at all times an attempt to gain insight into questions
whose answers appear unsatisfactory on the basis of existing knowledge; at the
beginning of all theory is the question. Theory is nourished by the unfinished, the
yet to be found is always inherent in it. Thus, however, it is also inherent in the
peculiarity of theory that it always goes beyond that empiricism which, as systematically recorded, is considered the only permissible one for science. It is precisely
out of this unfinished relationship that Theodor W. Adorno could say:

> But it is peculiar to theoretical drafts that they do not coincide with the findings of
> research; that they expose themselves to them, venture too far ahead, or, according to
> the language of social research, tend to false generalizations (…). However, without
> speculation's venturing too far, without the unavoidable moment of untruth in theory,
> it would not be possible at all; it would reduce itself to a mere abbreviation of the
> facts, which it would thus leave incomprehensible, pre-scientific in the proper sense.
> (Adorno 1998, p. 101)

Having made these purposefully general preliminary remarks, it is now possible to
summarize: Sociology is dealing with a fundamental question: How is social order
possible, how does it develop, how does it change, and how does it stabilize? An
answer, according to current understanding, is not possible without a theory of
empirical subjects. "Regardless of the changing trends in sociological debates,

therefore, the connection between subject theory, socialization theory, and social theory remains a central desideratum of sociology" (Sutter 2003, p. 47). Nor should it be overlooked that we are dealing here with the old, never dismissed question of the contradiction between coercion and autonomy, between individualism and collectivism, by which sociology has been driven since its beginnings. But what problems does it face? First of all, it can be stated: The contemporary world seems to have gone so far out of joint, or at least so far away from old ideas of order, that there is no longer any society as a unified social entity into which people could be socialized (Tilmann Sutter). Thus, even a general theory of society gets into shallows. Parts of the theoretical discussion suggest that the functional differentiation of modern societies could reach a level at which both the conventional delimitations of societies and the models used in research (because of their underdifferentiation) could become obsolete (Schimank 2000). On the other hand, the debate has indeed broken away from the conceptions of socialisation research, some of which are more than thirty years old, and has discovered, for example, the idea of "self-socialisation" (e.g. Zinnecker 2000), which emphasises the constructive self-activity of forming subjects. But the coupling with theories of functional differentiation does not seem to have succeeded yet. It also seems to me that old wine in new bottles is on the way here. Helmuth Plessner already spoke of "autonomous self-transformation" and Arnold Gehlen of "independent production". What can be expected from such coinages of words is that the meaning behind them or in them is aligned with what can actually be understood as changed social conditions of subject formation. The ethnic and cultural transformations, the rapid spread and use of new electronic media (especially the so-called social) and other phenomena could be made fruitful here in the course of empirical research. If this succeeded, research would be a step further, but another problem would still be unsolved: the special attention to the mediating relationship between man and environment; it already emerged with Jean Piaget, who, on the basis of a genetic epistemology, conceived this relationship as accommodation and assimilation. Jean Piaget's ingenious idea consists in a concept of learning that is based on the acquisition of schemata that are acquired under the "pressure" of the environment and at the same time are constantly autonomously differentiated and reshaped. Whether "autonomous self-transformation", "independent production", "biological-cognitive adaptation" or "self-socialisation" – in all conceptions, a force, a faculty, an activity tendency of the organism, which drives the subject and which is nowhere satisfactorily argued out, works in the background.[1] Even the biochemical processes in the

---

[1] Jean Piaget calls it "the main characteristic of the organization of life", which does not make things any easier.

cell have a motor in the background. A perusal of all the elements belonging to the definition of life in the most diverse Internet sources testifies, to my perception, to the implicit assumption, quasi pre-theoretically, that there is such a moving force, without being able to name it. Even older concepts of an ontologically understood biology of aging still have an effect by assuming an arc of development (of which Thales of Miletus already knew) from the build-up via an optimum to the decline of the life process, which is also driven by such an anonymous force (Kment 1996). Interestingly, there is a connection from this model conception to views of recent biology, which understand aging as an irreversible process in the guise of irreversible individual processes in a highly complex feedback system (organism). I see a possible path to an answer, albeit in a specific epistemological perspective, in Günter Dux's concept of self-care, which I related back to the recursivity of organic processes above. I am aware that this represents a considerable volte-face, for in neurobiology, too, recursivity is ultimately a construct.

The malaise of which I have spoken here thus makes it difficult to derive a secure concept of participation on the theoretical level of a sociological understanding of the individual and society. It is therefore necessary to tie it back to the definition in the first chapter and at the same time to enrich it on the basis of empirical research results. This is now done, following the plan of the considerations, above all with regard to age.

3.2   Age Specificity[2]

3.2.1   Age Structure Change

It is likely that in a retrospective that could be undertaken forty years from now, the ageing of the population on a global scale would turn out to be the most powerful factor in social and cultural change. Already in the middle of the twentieth century, a process of social transformation began to take effect on an intensified scale, which had already made its first harbingers visible in the nineteenth century, and which today is referred to by the notion of global demographic ageing, "age structure change" according to Hans-P. Tews (1993). In a rough perspective, this is a matter of an almost secular restructuring of the quantitative relationships between the various age groups, which entails qualitative changes, not least of which are altered forms of participation in the life cycle as a result of the socialisation

---

[2] Part of this chapter represents the heavily revised and abridged version of a text by Franz Kolland and Anton Amann (2013).

processes of old age. Although this topic of ageing has meanwhile been turned through all the mills of problem authentication, it still has something of the magic of the incalculable about it. Current population projections predict a growing number and proportion of older people in most countries. That the ratios between different demographic characteristics vary tremendously around the world is part of this puzzle. But the consequences for society are difficult to grasp, are often seen in a one-sided way and are reinterpreted as catastrophic, which is why the statements also oscillate enormously. When someone is old or is classified as old depends on physical, psychological and social factors, whereby social here also includes economic and cultural dimensions. Since these are mostly attributions, the range of informal but nevertheless behaviourally effective age demarcations is legion. At the societal level, or in the area of legal regulations, there is a clear demarcation in connection with participation in gainful employment, namely the legally fixed pension age, which has its own history of constant shifts since the end of the nineteenth century; it excludes many people from that life-centered participation which has come to be called economic. No corresponding demarcation is found at the psychic level. Here we have to do more clearly with changes which are to be classified as disease-related deviations, but which frequently do not occur in an age-associated way. This will be discussed later in connection with disorders and diseases in old age. When it comes to the question of when someone classifies themselves as old, or at what point someone is considered old, it becomes very clear that the answer depends on one's own position in the life course, on social integration, i.e. on the yield of participation. The older someone is, the higher the threshold is subjectively set at which someone is classified as old; interestingly, it is precisely this conditional relationship that is important for variations in participation behaviour. In biological and medical terms, too, there are probably no clear-cut demarcations. The research literature points to a variety of what are known there as biomarkers of aging, that is, biologically determinable objective indicators of disease, irrespective of any particular modality of their recording (Hampel and Pantel 2008), whether that is handgrip strength, physical mobility, oxidative stress, or lung function. What all these approaches have in common is that they see ageing as a multidimensional process which, given biological changes, is strongly influenced psychologically and socially (cf. also Amann 2008a). If we talk about age and ageing, then we need not only a view of the biological and psychological ageing process itself, but also a view of the social situation, gender relations and ethnicity, characteristics which all find their way into the so-called socialisation of old age. Through this broadening of perspective, the social heterogeneity in old age becomes visible and therefore generalized images of old age that refer only to physical changes lose their persuasive power. In such a broadened perspective, there is talk of "developmental gains". However, they presuppose that goals are determined and occupied by corresponding motivations. We are shaped by our backgrounds socially, professionally and culturally, and this is massively expressed in late life. Sigmund Freud was still so impressed by the power of origins that he no longer began analysis with patients over 50. What was then only seen by his "dissident" Alfred Adler and adopted in his concept of individual psychology was: we are also shaped by the goals we set or affirm ourselves, an idea that in more recent research led to the idea of the plasticity of age (cf. Amann et al. 2010a).

3.2.2 What Does "Ageing" Mean?

The demographic ageing of a society in the sense of an increasing proportion of older people is determined by three factors: Birth rate, life expectancy and migration. The most important causes of population ageing are the decline in infant and child mortality (rising average life expectancy at birth) and, in parallel, the decline in the average number of children per woman. Both measures are declining almost worldwide, and have been declining for decades in more developed societies. So what is an ageing society? In the general discussion, too little care is usually taken to distinguish between the processes of individual ageing and questions about the ageing of the population. Both are important for the developmental dynamics of society, albeit at different levels. This is closely related to the fact that people are increasingly talking about the ageing of societies. However, societies do not age in the sense usually assumed here, but people and populations grow older; their ageing, however, has massive consequences for societies, as can currently be observed in China, for example. Three perspectives on age(ing) need to be considered:

1. Social gerontology is a transdisciplinary science because old age and ageing encompass physical, mental, and social issues, and thus can only be meaningfully studied in complex, interdisciplinary problem settings. This notion has always prompted calls for interdisciplinary research (cf. Amann and Majce 2005).
2. Age and ageing are historical and ontogenetic (the specific development of the individual), so there are vast differences in the process and outcome of ageing between individuals, between genders, between cultures, and between historical phases.
3. Age, and especially old age, is an area of societies that is still poorly shaped, because although there have always been old or very old people, a steadily increasing probability for more and more people of being able to reach a high age

is only an achievement of the twentieth century. A theory of the socialization of age with an emphasis on old age is a desideratum of social gerontology.

From this we can deduce: An ageing society is not simply a society in which there are many older people. There are several ways of determining this, one of which is to assume a ratio: For example, if the proportion of people under 20 is smaller than the proportion of people over 65. What are the implications of such an "ageing world"? Public discussion is very often determined less by factual debate than by arguments pointing to fear of old age, just as in recent years playing on fear has become a favourite pastime of governments and the media, who could be accused of lacking sanity. First, there is the fear that the existing health care systems will not be able to cope with the growing number of older people in need of care. The counter-argument here is the thesis of the compression of morbidity (Fries 1983), which assumes a relatively decreasing need for care. Secondly, an economic collapse is expected as a consequence of the unfavourable ratio between the employed and the retired. However, this fear seems to be subject to seasonal fluctuations, depending on changing short-term economic growth forecasts. A rational approach to this issue can only be expected when it is understood that a completely new concept of productivity must be assumed in this context. As a consequence of this presumption of strain, a generational conflict between the working population and the elderly is envisaged. A solution is seen here in remaining longer in gainful employment and in learning, which should accompany the whole life. It is not without reason that participation in educational processes in old age is one of the most well-trodden fields of research. A third danger is perceived in the structural change of the family. Weaker emotional family ties are suspected and thus an undersupply of the older family members. These fears are countered by the objection that there is a high potential for intergenerational solidarity in society. Finally, there are fears of economic stagnation as a consequence of an ageing society. On the other hand, it is argued that the health and social sectors are developing very productively in ageing societies. This whole discussion also belongs to the topic of the so-called images of old age and their fluctuations in emphasis, which will be discussed in more detail later. With reference to the third point mentioned above, high age, there are some empirically significant findings which indicate that involvement and disengagement in old age are almost necessarily reshaped once again. There is a striking heterogeneity with large differences in the extent to which people are affected by age-related limitations and increasing differentiation in old age; nevertheless, there are relatively large groups of men and women with comparatively good health, a high degree of independence and autonomous living, which is in stark contrast to the prevailing and predominantly deficit-oriented image of old age (association of old age with illness and need for care) in society. Almost half of the very old are also affected by more or less pronounced frailty; frailty is often accompanied by multimorbidity, increasing mobility restrictions and limitations in the ability to help themselves. Especially between the ages of 80 and 85, a significant increase in age-related functional limitations begins, combined with an increase in the need for help and support as well as the need for long-term care. Although women make up the largest proportion of the very old population, they tend to be in poorer general health than very old men. They are predominantly more affected by chronic diseases, suffer more frequently from pronounced mobility restrictions and are more frequently dependent on support or care. The higher their education and income, the healthier the very old are. It is highly likely that people with a lower socioeconomic status are also less likely to reach old age (cf. Ruppe and Stückler 2015, pp. 15–17).

3.2.3 Tasks of a Political-Practical Nature

When talking about ageing societies, at least at present the sustainability of social security, i.e. of the health, pension and care systems, is at the centre of public discussion; questions and realities of the kind just mentioned are notoriously neglected in the public debate. The prevailing discourse is part of the wider welfare state critique that has begun to dominate thinking since the 1970s in the wake of neoliberalism. Most macroeconomic estimates suggest that total age-related public expenditure will increase over the next few decades. In this context, there is talk of a (fiscal) sustainability gap. Although spending on schooling and unemployment benefits might decrease due to the declining school/study and working-age population, respectively, spending on pensions, health and long-term care will increase significantly due to population ageing, leading to an overall increase in public spending.[3] The increase in public expenditure could then be less if the health situation of the old and very old people improves significantly. If people live longer in good health, expenditure on health and long-term care will fall and the sustainabil-

---

[3] The fact that there are constant changes in the statistics here does not make the picture any simpler. In Austria, for example, the increase in the de facto retirement age was recently highlighted in a positive light. However, part of this increase is artificial, a statistical artefact. With the exclusion of invalidity, which is reasonable in itself, about 20,000 rehabilitation money recipients were removed from the pension statistics, although they continue to be provided for from the same public funds. Moreover, the ever-present jubilant cries about a few percentage points of so-called improvement are nothing more than misunderstood numbers under petty short-term perspectives.

ity gap will close. The activation strategies in the way of targeted participation are particularly conspicuous in this area; people are virtually put under an obligation to be active. This is particularly evident in the attempts to put older people in the place of regular employees through volunteering and voluntary work. The leisure industry is also profiting considerably from this trend. In any case, it should be noted: So far, most activation programs, which are quite well-intentioned, start with individual skill maintenance and skill promotion. The effectiveness of such interventions is no longer undisputed. There are doubts that more activity leads to greater social interaction. This is because activation programmes can create and reinforce dependency and marginalisation, which they have been started to eliminate. Activation in residential and nursing homes for the elderly is often aimed at satisfying selective needs rather than at structurally changing environmental conditions. It partly takes place in a retreat of irrelevant social and societal activities and roles. Volunteer work, too, must go beyond activism if it is to be individually and socially successful.

Due to the disproportionate (in relative and absolute terms) growth in the number of very old people, a rapidly growing need for care and nursing on an unprecedented scale is to be expected. The rates of need for care and nursing are quite low up to the age of about 75. Then, however, this rate rises rapidly, reaching just under 20% for those aged 80–85 years and 43% for those aged 85 and over. In this context, we can thus speak of a second gap, namely a care gap. Studies on the baby boom generation in the USA foresee a widening of the help/care gap in the future. According to these studies, 15–20% of baby boomers cannot count on any family support when they need help/care. The gap is even larger for those without a spouse (Ryan et al. 2012, p. 185). Although the statements in this regard are highly speculative for very old people in need of care in 2030, a sharp decline in informal (private home) care is nevertheless assumed. In recent gerontological research, however, this care gap is controversial. Empirical findings are put forward that point to a widening, i.e. rising care rates linked to demographic ageing, as well as evidence that points to an increase in active and disability-free years in this increased life expectancy and thus to a decline in the need for care. The latter position is associated with James Fries (1983) (see above), who developed the "compression of morbidity" thesis. Until his work, it had long been disputed whether it was possible to postpone medically incurable chronic impairments, as well as physiological ageing processes, to such an extent that they were compressed to a few years before death. His thesis was apparently understood and accepted by only a few. The idea was (and still is) widespread that with increasing longevity one must reckon with overall poorer health and frailty in the elderly population. Epidemiological data, however, speak a different language in some cases: although there are high numbers of multiple and chronic diseases among the very old, by no means all of them require permanent treatment, and the number of diagnoses is in any case considerably higher than the number of therapies used. However, long-term observations suggest a compression of the burden of disease and disability. Empirical studies from the USA (e.g. Manton et al. 1998) show that the actual course of chronic disabilities in older Americans is lower than the predicted course would lead one to expect. A starting point for future health and mobility in old age is seen in the prevention of chronic diseases. Overall, the results and deductions from these observations remain controversial in their long-term significance. More recent studies see compression as a transitional phenomenon in modern societies and assume an increase in morbidity in old age (cf. Boongarts 2005).

However, the discussion about an ageing society is not only a question of economic and health considerations. It is also a question of the design of the central social institutions (family, education, gainful employment) and the normative regulations linked to them, which decide on integration and disintegration. Age is socially unspecifically formulated, so that one can speak of a third gap, namely a normative gap. Normative gaps can be found, for example, with regard to the employment of older people. Even though the Charter of Fundamental Rights of the European Union contains a ban on age discrimination, people who become unemployed after the age of 50, for example, have considerable difficulties in finding employment again. But it is not only in the world of work that normative changes are needed because of longevity. Normative changes are also needed in health and care settings and in educational and recreational settings. To close the normative gap, the UN General Assembly adopted the strategy of mainstreaming ageing in 2002. This strategy pursues the goal of integrating all aspects of ageing into all relevant policy areas at all levels. This is intended to stimulate normative changes in the political system, in the health and education system, in church-religious institutions, in the world of work and in the family, in order to do justice to the newly created life phase of old age in all its dynamics.

3.2.4   Ageing Is Changing

Everyone is repeating it to themselves: The life phase of old age is undergoing change. Since the 1970s, social gerontological studies have called into question not only the deficit model of ageing, but also the homogeneity of this phase of life. The essential change that is evident here in scientific thinking is one that can be summarised under the term differentiation. Differentiation refers to long-term changes in society that are associated with a new emergence and increased division of social positions, life situations and lifestyles. The causes of increasing social differentiation are the growing division of labour, longevity and the formation of diverse forms of life. Social differentiation thus describes the breakdown of a uniform whole and has the effect that individuals no longer form uniform identities and thus less homogeneous life situations arise. This has been discussed above in the topic of theories of society and socialisation.

Here it is worthwhile to take a small, selective look back. One of the first attempts to make these changes visible for the post-occupational phase of life was undertaken by Bernice Neugarten (1974) by distinguishing between the young old and the old-old, which is indeed a crude classification. To this end, she undertook an age-based localization by setting the young age between 55 and 75 years and the old age above 75 years. However, she herself admitted that such an indication was unsatisfactory because chronological age was not a reliable variable for describing social differentiation well, but it was nevertheless indispensable as a "boundary marker". Studies on the quality of life in old age have long since demonstrated that chronological age is not an explanatory variable. Neugarten described the young elderly as relatively healthy, affluent, free of traditional family obligations, and well-educated or politically active. This group, Bernice Neugarten argued, would produce an "age irrelevant" society via new needs and active management of their lives. The old old are those who need care and services due to health burdens. It is in this group that participation in life often becomes precarious. The idea of age irrelevance has, however, been undermined by development, for presumably no branch of the consumer-economy is now flourishing better than that of long-distance travel for the active old.

Another significant contribution to the description of the differentiation of old age was made by the English historian Peter Laslett (1995), who divided the old age phase into a third and a fourth age. This distinction not only provoked an enormous reaction in academic research, but also influenced the practice of service provision in old age. In retrospect, it can be stated once again: Categorizations create facts. So what is new about this conception? Whereas the third age is a life stage of choice, expanded opportunities, creativity and personal development, the fourth age is characterized by dependency and decline. The young old (third age) live largely free of disability, while in the very old (fourth age) age-related physical and mental-psychological limitations force adjustments in everyday life. Even if Peter Laslett did not want this classification to be tied to a specific age, in the social science discussion people over 80 years of age and especially those over 85 years of age are counted among the group of the very old. This definition is based on demographic considerations and figures on the prevalence of care. Peter Laslett's division into third and fourth ages completely detaches itself from calendrical age and replaces it with a concept based on standard of living and generational cycle. What was not taken further, however, in comparison with Bernice Neugarten, who had still described an age-irrelevant society as a vision of the future, was the overcoming of a binary view of age and a stronger linking of objective and subjective influences.

Does this put an end to the threefold division of the life course? According to Martin Kohli (1985), the development of modern industrial society is characterised by the formation of a regulated life course (cf. the considerations on life course policy below). In terms of the life course, this means that in modern society one can observe a life course that exhibits a comparatively high degree of age gradation. This process, described as the institutionalisation of the life course (Kohli 1985), refers not only to the orderly progression of the life course, in which, for example, the age limit has a strong regulating effect, but also to the mode of action itself. What is meant by this is an increasing biographisation of the life course, i.e. the life course is increasingly shaped by the individual and less influenced by family, class or religious affiliation (cf. Amann 2017). Especially in more developed societies, this aspect points to the possibility of an expansion of opportunities for participation. Part of the institutionalisation of the life course is the division into three phases: education, employment and retirement.

In the tripartite division, Mathilda and John Riley (2000) see a structural imbalance for the standard of living of older people. Although they expect to live longer and longer in better and better health, they are released from gainful employment into leisure roles via fixed age limits. In this context, the two researchers speak of a "structural mismatch", by which they mean that there is a gap between the existing skills and potential of old age and the roles that are actually available. Apart from the grandparent role, there are hardly any other social roles available to older people. They refer to this social structure as "age-segregated," meaning that the education phase is followed by the employment and retirement phases in a linear sequence. This three-part division of the life course in its linear sequence must be dissolved in favour of an age-integrated structure, in such a way that all three phases of the life course no longer run consecutively, but simultaneously. Age-differentiated structures not only have the disadvantage that they do not fully exploit the potential of old age, but also generate forms of social exclusion and segregation.

While the model presented by Mathilda and John Riley is primarily concerned with the dissolution of the tripartite division of the life course, which is seen as an obstacle to the unfolding of a productive old age, a more recent model of the life phase of old age by Miwako Kidahashi and Ronald J. Manheimer (2009) assumes a "post-retirement society". While the twentieth century was dominated by the model of retirement after a long period of gainful employment, the twenty-first century is seen as one that is transforming through changing gainful employment, lifelong learning, and expanded leisure into an age in which the institution of retirement as a life stage of "resting" will disappear. Not unlike in other fields of research, the categorizations and guiding concepts in social gerontology are constantly changing; only the constant empirical control of meaning is ultimately able to determine the durability of assumptions.

3.2.5   Sensible Pragmatism

From a pragmatic perspective and born out of a description of reality that sees both the economy and the welfare state in crisis, social gerontological research in recent years has concentrated more than ever before on questions and problems that concern the standard of living of older and very old people and are directly related to the preconditions and consequences of the socio-political shaping of social living conditions. Demographic change and structural change in ageing are consequences or results of the modernisation of society. In their interaction with elements of modernisation, they are the background for the developing need for a new socialisation of ageing. Socialisation refers to material and normative integration into society. Depending on the society, there are different means and goals of integration. Successful socialisation is achieved when the individual has an acceptable livelihood, meaningful and recognised employment, satisfying social contacts and satisfactory health-related facilities.

It is not demographic development and age structure change – as is often assumed in the public debate – that are the cause of the hitherto unsolved challenge to society. Ageing cannot be adequately understood if it is only seen as a problem for society. It is only in connection with the processes of change in the economy, politics and culture, in the labour market and with regard to social policy and the family that demographic development and age structure change also contribute to the fact that the hitherto institutionalised socialisation of ageing is becoming increasingly problematic. The previous forms of socialisation relevant to ageing are not dynamic and flexible enough to keep pace with overall social development. Meaningful pragmatism would have to be linked to the preconditions that I have outlined above on the basis of the new conception of generational policy.

3.2.6 Malfunctions and Restrictions

Surveys in various European countries (cf. SIZE 2006)[4] have shown that, from a social science perspective, the concept of mobility is suitable for placing research findings in a systematic context when considering disorders and limitations that are largely associated with ageing. First, some conceptual clarifications are useful. When speaking of mobility, the mental reference point is usually a movement in space or a spatio-temporal process. This is also the common understanding: mobility means the movement of people and goods in space in order to overcome distances. However, this concept, which is based on the idea of traffic, is too narrow for our question. Now, although the term mobility may have originated in eighteenth century military language, it has been used in psychological and social contexts from the very beginning. The German geographer Jürgen Bähr has therefore used a general systems-theoretical paraphrase: Mobility means the change of an individual's position between defined units of a system (Bähr 1983, p. 278). In addition to these conceptions, the term carries the notion of a cognitive faculty in the sense of promptness and responsiveness of thought (mobilitas animi). This mental mobility has always been seen as a correlate to physical mobility. Meanwhile, it has become common to distinguish between social and spatial mobility, primarily targeting the willingness and actual realization to change one's place of residence as well as social networks and educational and work environments. These examples show that the concept of mobility is understood in very different ways. For the purpose of the present research question, I will try to give it a specific definition.

In principle, an individual must be able to move in space; if he or she does not have this ability, participation in life is not possible without support. It can therefore also be said that mobility constitutes the experience and use of living space (Topp 2001). Although this prerequisite presents itself as indispensable for participation in life, there are considerable interindividual degrees or extents of mobility that are considered necessary, and there are highly diverse goals in the service of which mobility is placed. In the literature, goal-directed mobility has been associated with needs whose satisfaction may conflict with the physical and social mobility conditions of the environment. Social gerontology has adopted a transactional notion of mobility, clearly expressed in the "Environmental Docility Hypothesis" (Lawton 1990). The individual and his or her environment are reciprocally connected. Now, as someone grows older and as losses and deficits increasingly

---

[4] Besides Anton Amann and Barbara Reiterer, Ralf Risser (FACTUM, Vienna) and Heinz-Jürgen Kaiser (University of Erlangen) were involved in the compilation of the report cited here.

emerge, the specific configuration of the environment becomes more and more significant for the planning and realization of everyday life and vice versa, or in other words: the conditions for involvement become more difficult without adaptation of the environment to the changed needs of the individual and without adaptation of the individual to the environment.

It is now necessary to address an aspect of the mobility debate that has become increasingly ideological in recent years, in that the discourse itself has become normative. In contemporary society, mobility has long since taken on a positive connotation; completely divorced from individual needs and intentions, it is seen as synonymous with active, dynamic, vital, flexible, industrious and adaptable. Mobility has become a slogan for a good life and thus a call for a certain lifestyle. This in turn has to do with the fact that participation of older people (called social participation) is on the social and political agenda because, quite rightly, it is hoped that participation will prevent disengagement and exclusion. It becomes difficult and costly to organise adequate support for the elderly if they are cut off from social institutions and services or isolated from them.

Increasing age is associated with a range of physical and psycho-mental limitations and deficits. Sensory, cognitive, mental and organic abilities can become limited, which inevitably leads to changes in mobility, which in turn can result in a loss of quality of life and increased cost burdens. For at least a quarter of a century[5] some empirical relationships have been repeatedly confirmed. A comparison of mobility parameters among older people shows that in 50% of the days in a year they do not leave their home (agreement for Belgium, the Netherlands, the United Kingdom and Switzerland, see SIZE 2006, p. 19). The percentage depends, among other factors, on chronological age and gender. For those aged 75 and over, the figure was 60%, and for women of this age, 70%. Causes vary; some research indicates that while there is a discernible decline in the amount of time older people spend away from home compared with younger people, there is no discernible decline in the number of out-of-home stays (Hartenstein and Weich 1993, p. 39). The personally reported motives include not only physical limitations but also attitudes and emotions, especially the fear of becoming a victim of assault on the street.

The range of movement narrows with increasing age. In this case of reduced mobility, attention should be paid to barriers that may be unnecessary and could be removed.

---

[5] For a comparison of the data, see the older sources cited in the text and the peer-reviewed findings in Amann (2011).

3.2.7   Mental-Psychological Changes

For our topic of the conditions for participation in life, it may suffice to speak of psycho-mental changes in old age and of mental disorders and diseases in old age, even if this division may seem somewhat crude (cf. Jenny 1996).

The concept of a general decline in mental performance in old age, a form of change in intelligence, has a long history. The so-called "deficit hypothesis", the assumption that there is a successive reduction in mental performance from the third decade of life (Lehr 1991), has long since been refuted. Early on, Leopold Rosenmayr (1974) treated the subject critically in a comprehensive literature review. Nevertheless, this view still prevails, especially for old age as a whole phase of life; not only among younger people, but also many older people perceive themselves in this way. In this attitude, for it should not be anything else, a generally widespread, tendentially pejorative attitude towards old age comes through – contrary to all the euphoria about old age that is spread by advertising.[6] In fact, however, not all people undergo the same changes at an advanced age, so that a variety of interindividual possibilities of mental changes and also possibilities of control occur: Degradation and compensation, selection, optimization, and activation and exercise (see Amann et al. 2010a, Sect. 5.4). However, we know little about whether and how attentive observation of changes and targeted counteracting activity can increase vital energy, and not only delay a further decline assumed as a model. Relevant preliminary considerations were published by Erik H. Erikson, Joan M. Erikson, and Helen Q. Kivnick (1986). This question, however adequately framed in terms of research, might touch on the point at which increased energy/ activity levels are likely to increase the extent of participation. Leaving aside the decline in age-related speed in intellectual performance, the differences in average performance among healthy members of the same generation between the fifth and eighth decades of life are relatively small.[7] Age plays a subordinate role in all of these questions because no uniform deficit pattern of age-related changes in intelligence has yet been empirically demonstrated. Environmental factors and health status, on the other hand, play a significant role. Level of schooling and continued intellectual activity have an empirically proven broadband effect into old age. The

---

[6] The grandmother suffering from a degenerative, immobilizing and painful joint disease can suddenly romp through the garden with the dog and grandson in the morning, if she has only rubbed her knee with the supposedly best ointment available on the market the night before.

[7] I will not go into details here, such as the distinction between fluid and crystallized intelligence and what might be genetically determined about the former; they can be found in any review of the psychology of ageing.

state of health becomes subjectively as well as objectively more significant with increasing age, e.g. age-related sensory losses have an effect on intellectual performance. In addition to brain-organic changes, which will be discussed later, various other organic diseases, which occur with increased frequency in old age, can influence intellectual performance.

Although the research results, especially the theoretical models or concepts that have been developed in recent decades, appear to be very diverse and relatively heterogeneous, we will nevertheless briefly discuss memory changes in old age. "Memory", a faculty of the individual that is partly interpreted as located in the brain, partly as a modal activity of ego and self, is a cognitive performance domain that changes significantly in older age in the vast majority of people. We notice changes in memory and recall most quickly in everyday life, and therefore it is to them that we attach notions of cognitive performance. The changes come insidiously and almost unnoticed. When personal names and dates are no longer remembered, we still help ourselves with tricks. There was that writer, an Austrian, what was his name? He wrote something about a "Hagestolz", lived at the time of Abraham Lincoln, what was his name? He committed suicide. He also wrote A Midsummer Night's Dream. No, it's After Summer. Is that the right title? Ah, yes. Adalbert Stifter. Now, cognitive performance can be diminished by psychological stress as well as organic impairment, but this can occur not only in the elderly but also in the young. It is fascinating that the same phenomenon is interpreted differently in younger people than in older people, in whom it is generally regarded as an age-related deficit. But again, age-related functional deficits are only one cause in a whole series of factors, and training and memory strategies are effective.

Personality changes are a little-known and usually misperceived phenomenon, especially in everyday life. To some, they appear almost mysterious and fateful. This ranges from the "stubbornness of old age", an inconsistency of opinion which is difficult to correct and which is also present in younger people, to emotional and mental weaknesses which appear to be due to a change in attitude and which are used to misjudge the onset of depression, to the idea that psychological reactions are due to biologically induced processes of decomposition (Jenny 1996, p. 61). It is precisely here that the individual-environment entanglement, which has already been explained several times, acquires eminent significance. Instead of perceiving conditions that make coping difficult or hinder it as such, the problem situation is individualized. This then leads to assumed age-related personality changes, which are interpreted as increasing introversion, decreasing flexibility and increased depressive reactions to stress, and which are almost entirely without scientific foundation (Jenny 1996, p. 62). One can hardly go wrong with the assumption that the time characteristics and environments of a generation have a specific influence on personality development. It is not for nothing that the generation of writers after the Second World War (e.g. Wolfgang Borchert or Heinrich Böll, the so-called "Trümmerliteratur") was preoccupied to such a conspicuous degree in their works with war experiences, poverty, hardship and missed opportunities in life. It is equally true that certain so-called personality traits turn out to be both stable and changeable over the life course. Openness and restraint, flexibility or rigidity, which people who have grown old show, also characterized them in their younger years. The level of activity and mood in the sense of a "precursor" of depressive reactions also hardly change. Creative abilities and artistic potential remain relatively stable into old age. Age hardly shows any influence on these personality traits (Jenny 1996, p. 62).

One of the mental illnesses that has received increased but still insufficient attention in recent years is dementia. A quarter of a century ago, scientific discourse was still strongly medically and biologically oriented (e.g. Bergener and Finkel 1995), but soon the spectrum of attention became broader and the social sciences (psychology, social work science and sociology) also took up the issues (e.g. Aldebert 2006). It is estimated that 13,000 people in Austria are living with some form of dementia (Höfler et al. 2015). Due to the increase in the number of older people, especially the very old, the proportion of people with the disease will double by 2050. The debate about possible causes of dementia has been going on for decades (cf. Pantel 2017), and in the USA the concept of dementia as a syndrome diagnosis has now been abandoned in favour of the concept of minor or major neurocognitive disorder (NCD). Symptomatology and disease progression are not to be discussed here, but a consideration from a sociological perspective is unavoidable. In the foreground is the perception that diagnosis is difficult and often inaccurate and that there is no effective therapy. The disease seems to be accompanied by a significantly reduced life expectancy, it manifests itself in severe mental and physical impairments, it massively worsens the quality of life and in its progression becomes an ever increasing burden also for the environment, primarily for the caring relatives. Those affected by dementia and their relatives experience isolation and exclusion from social life, they are pushed out of the public sphere and step by step removed from their environment. It is a suffering of exclusion, of deportation (Schaub and Lützau-Hohlbein 2017), which requires sensitivity, care and sound training in those who deal professionally with the situation, but this is not always guaranteed. How society views dementia is associated with or dependent on specific patterns of perception. In earlier times, this perception was dominated by images of the disease in its middle to late stages; currently, attempts are made to provide support at early stages and to focus more on possibilities than on deficits. In the practice of classification, the concept of independence therefore plays an increasingly important role. To reiterate: Dementia is the most common and at the same time the most serious mental illness; it not only leads to the gradual removal of the affected people from their life contexts, it also makes it massively difficult for the relatives to participate in life.

3.3   Generations Are Interdependent

There are two reasons in particular that suggest the inclusion of the generation theme. On the one hand, the concept of generation fits logically into the connection between the individual with his or her acquisition of competence in action, language and thought on the one hand, and the others in the everyday world, represented above all by preceding generic members; on the other hand, the succession of generations represents a core of the demographic structure of a society. These connections will be briefly examined here; the discussion establishes a reference to Hannah Arendt (1981) and Hans Jonas (1980). In both, the idea of being born and leaving is related to each other, whereby Hannah Arendt sees the particularity in the regrowth through action and language of those who come, while Hans Jonas sees it in the beginningness and otherness of those who come in each case; their arguments are surprisingly similar. This idea was probably already a topos in the Biedermeier period. Adalbert Stifter says in the juxtaposition of a youth and an old man ("Der Hagestolz", cf. Sect. 6.2): "Victor the free cheerful beginning, with gentle flashes of the eyes, an open place for future deeds and joys – the other the degenerate, with the intimidating look and with a bitter past in every train" (Stifter 1951, I: 638). That we must die, says Hannah Arendt, is linked to being born. This places youth and old age in a relation of imitating and departing. "Because every human being, by virtue of being born, is an *initium,* a beginning and newcomer to the world, human beings can take initiative, become beginners, and set new things in motion" (Arendt 1981, p. 166). This is how it has always been ordered. The changes of the last hundred years cause us to reconsider the meaning of this connection in the space of actual developments. Birth[8] and mortality are linked to the promise of beginning, immediacy, and the eagerness of youth, along with a constant supply of otherness that emulates. There is no substitute for this in our world. The philosopher Hans Jonas, for example, thought this far decades ago. "This always-again-beginning, which can only be had at the price of always-again-ending, may very well be the hope of humanity, its protection from sinking into boredom

---

[8] Hannah Arendt speaks in English of natality; the expression Gebürtlichkeit chosen for it in the German translation sounds all too wooden (Arendt 1981, p. 167).

and routine, its chance to preserve the spontaneity of life" (Jonas 1980, pp. 49 and 50). But the counterpart has not yet been seen: that departure also involves promise. Old age provides those who come after it with the model of how they either can, or will not, or should become themselves. All who grow older have their mirror in those who have already grown old. From it come their fears and their hopes, their practices and their ideologies. There is no substitute for that in the world either. Children go to parents in opposition, they detach themselves from them. Without this process, they could not "grow up." On a large scale, the succeeding generations are always new. They have grown up in different times, they make up their own world. But never without looking at what has gone before them. In the process, the concrete forms change. Once upon a time the ancients may have had authority and wisdom and the young may have learned from them. Today they have largely lost respect and authority. The "reverence for snow-white hair" was already only a fiction in the pop songs of fifty years ago. Today, it is said, the old learn from the young. But this learning most likely refers to the new techniques of coping with everyday life and to information technologies. Whether the young learn from the old, both from each other, or the old from the young, are manifestations of historical transformations. So far, too little attention has been paid to the mental and spiritual dimensions of generational learning (Amann 2004). Yet one dimension of the generational relationship is also of ethical significance. The self-blame denounced by Immanuel Kant in his Enlightenment essay is also a generational relationship and doom: each generation educates its children to total or partial immaturity by imposing and introducing them to its own, itself insufficiently clarified, understanding of the world, its use of language and concepts in an unreflected manner. Elsewhere (Amann 2017) I have noted that it is always the adults who already inculcate the prejudiced, derogatory and condemning stereotypes in children, that it is their crime, not legal but moral, to deform consciousness and character (without exception, whether religious or political). What will those today's boys have learned from their parents who are already back to basking in political leader worship, mass media dumbing down and who "grundeln" in dull nationalism?[9] It would certainly have been an interesting political science project idea, about thirty years ago, because at that time this whole absurdity began to show its first traits, e.g. with the Austrian politician Jörg Haider, to record the change in political strategies and the acceptance in the population in a long-term observation. What would be understood by strategies? Those patterns of action that

---

[9] An expression from the Viennese, which denotes the completely inattentive swimming in the murky shallows, to which I cannot find a more appropriate word, which in its latent content refers at the same time to the character and the circumstances.

have emerged among the populist right. These include constant provocation, which pushes the boundaries of language and moral sensibilities solely in order to gain public attention. These provocations insinuate that the right is engaged in a constant righteous struggle against powers that threaten the well-being and security of the people (asylum seekers, welfare parasites, etc.). When they are criticized by other parties for their provocations, they portray themselves as victims (treacherous enough is the origin of their buzzwords like "lying press", "asylum tourists" or "traitors to the people"). In the process, a dialectic precisely calculated by them plays itself out. If other parties are also provocative, the right's offensive counterattack benefits their attention (their sympathisers see them confirmed in their victim role); if they deliberately hold back, their profile is too blurred, which many again see as confirmation of the correctness of the right's arguments (in Austria, the SPÖ has been in this precarious situation for about a year). If the right were to behave in a "serious" manner, it would probably suddenly become less interesting for a large part of its supporters. But it must not exaggerate, too great is the danger of exposing itself. The best evidence for this is that in the numerous verbal attacks by the far right, who openly use threats of violence, slander and defamation, the party leaders come out with softening messages such as: that was not meant that way, the person was misunderstood, it was deliberately misinterpreted (the issue is turned over and the person is accused who took pleasure in pointing out a grievance). Without carrying this general presumption further, the question can rightly be asked again, "What will the little ones and young ones of today have learned when they step on the stage of shaping public life twenty years from now?"

A generation is an imagined life context of people belonging to the same cohort, on the basis of which a set of characteristics more or less common to all can be described. These characteristics include the aforementioned shared experiences and values, skills and knowledge, but also commonly experienced events such as war, the emergence of television and then the Internet, the fall of political regimes, or the dominance of certain musical trends. The respective composition of all these characteristics, which are usually characteristic of a certain, relatively short period of history, constitutes the "uniqueness" of a generation. It is thus permissible to speak of a generation as being a shared culture and tradition, a shared constellation of emotions, attitudes, preferences and practices. However, it must be emphasized that generations are not exclusive, because the above characteristics also overlap individual generations. The Beatles or Frank Sinatra, the gradual decline of major political parties, profit-driven environmental depravation quite obviously span multiple generations (Amann et al. 2016, pp. 37 and 38).

In an analytical distinction, three views of the generation concept have emerged in recent years that could develop relevance especially in empirical research. The sequence of generations in the family, filiation, focuses on micro-social networks with their tensions (see below) and practices of coping with everyday life, with their economic strategies and fictions of tradition. This perspective corresponds to the expression of generational relations. A next perspective focuses on relations of extra-familial processes between cohorts (age groups born at different times), it should capture time-historical and cultural exchange processes between different ages. Conflict and reconciliation in this concept do not take place through interpersonal relations, but through the mediation of institutions. Here, it is right to speak of generational relations. The third perspective is directed, as it were, towards a special form of the second, in that age-related large groups come into view, as in the so-called intergenerational contract. These large groups, e.g. the employed and the retired, are not adversaries who determine their circumstances more or less freely; these are controlled and defined for the most part by state institutions with the help of the economy using the instrument of redistribution. However, the following applies to all three forms of generation: all generations that grow up also have before them the material world of those who were and still are before them. In recent years, the argument has emerged that the old left the young a devastated and exploited world, that they acted without regard for the generations to come. Here, too, the view is too short. Those who do not look back do not understand anything. The degradation of the environment in its modern form has been going on for well over two hundred years. Almost all of them are harmful to the environment, and the major decisions that are detrimental to the environment are always made by those in full possession of economic and political power – these are never "the old people" at any given time. Here the idea needs a time dimension. What injuries and irreparable damage have been done to nature and human beings in our scientific-technical civilization is the responsibility of society as a whole in the long horizon of what has been and still is notoriously misunderstood as progress. This would be a burning occasion to reflect on an ethics of responsibility that reaches far ahead. The immovable core of all relationships, however, is the mutual interdependence. The young contribute to what can be called the constant reshaping of the world in their zeal for their future, the old by what they are, and to which the young orient their drafts – approving or rejecting. From this perspective, participation is any contribution of the different generations to the shaping of the world. That of the old is not less than that of the young. And to judge whether one is better or worse would first require socially binding assessment criteria in the light of a better society. These are not to be seen far and wide. If the side of the old is pushed aside for ideological reasons, as has become fashionable, the negation of the principle of being referred to one another occurs. Society is all or nothing. He who thinks only economically is a bad philosopher; he who decides in politics only technocratically has lost sight of the whole. But whoever allows himself to make judgments without having the whole in mind is not far away from charlatanry (Amann 2004).

3.4   Family, Relationships and Generational Solidarity[10]

There is hardly any other topic about which there is such a high degree of ignorance and misconception, and there is hardly any other topic about which there is such a one-sided focus in the public discussion as that of intergenerational relations and relationships. All the more important are the results of studies that provide empirical findings on attitudes and behaviour in the interrelationship between the generations or age groups, on the one hand in families, and on the other at the societal level. The central finding is that the family, in particular the intergenerational relationships of the nuclear family, continues to represent an extremely sustainable, solidary system of protection against emergencies and situations of need for help and support. If a need for support arises, hardly anyone is left without sufficient help from the family. Such a need for help occurs in part considerably more frequently among younger people than among older people. This becomes particularly clear in the financial area and in the care of children, and it is regularly the parents' generation that primarily covers this need for help. The central source of help is the mother. In the case of services for older people, the daughters also play an important role, just as helping in general has a strong female "flavour". The family is still a hub of solidarity between the generations. Mutual support takes place psychologically, socially and materially. The share of the elderly should not be underestimated: Their permanent material contributions to the younger generations are a significant element in maintaining their standard of living.

It is a fact that intergenerational relationships take place to a large extent within families. But it must also be borne in mind that "family" is not just family. There is a great discrepancy between the ideal conceptions of forms of cohabitation (marriage and family) and reality. According to research in European countries, an overwhelming proportion of adolescents and young adults imagine the family as the ideal form of cohabitation, but the reality is quite far from this ideal. In today's society there is a wide range of family manifestations with various types of relationships. These relationships are always connected with relationships of power, with dependencies, but also with the readiness to help, to assist and to put

---

[10] Here I refer to a not yet published text that I wrote for a research project. Cf. Ageing (2009, Sect. 3.4.2).

one's own interests aside. Often family relationships are exposed to stress, such as handicapped children, the care of elderly relatives, drug and alcohol problems, and psychologically difficult constellations such as mother-daughter or father-son conflicts. These difficulties often accumulate, for example, when it comes to the provision of care services. However, all these findings are not an expression of new conflicts; balance and conflict are inherent in all social relationships, including those between the generations in the family.

The relationships that can be observed become striking in the so-called middle generation, because this generation probably experiences the changes in the relationships to the children, their own generation and that of their parents most clearly. Studies show time and again that the birth of the first grandchild turns parents definitively into elders, i.e. grandparents, and some fiercely resist being called grandpa and grandma by their own children. One of the tasks that is not easy is that parents and children have to break away from the original parent-child relationship and establish one between older and younger adults, respecting in particular the independence of the younger ones. Many of the difficulties in farm transfers are also likely to arise from this context, not just from economic and regulatory constraints. Specific types of unsuccessful relationship work have long been known. If, for example, as often happens, the parents of the middle generation had too great expectations of their own lives and experienced their failure, they often look to the children to fulfil their narcissistic desires. As a result, children who have experienced such demands see themselves as being constricted for many years and have to develop in a certain direction under pressure. This image then includes the early leaving of the parental home under protest or they go into opposition within the family. Then occurring violent reproaches because of the "otherness" of the children reflect pronounced disappointments in the parents. In the novel "Die Nähe der Sonne" by the Austrian writer Gernot Wolfgruber, in which the protagonist Stefan Zell ends up psychologically destroyed, such a never successful relationship between Zell and his parents has been traced in an oppressive way: "And inside, inside Hiroshima." Another form of disturbed relationship is visible, for example, when the middle generation breaks off relations with the older generation because of jealousy towards their own children. All their lives they have longed in vain for affection, warmth and reconciliation. Now they experience that their own parents as grandparents grant all this to the grandchildren. Not far from this relationship problem is the case of caring women who patronise and dominate their old mothers in need of care, as if they wanted to "get back" at them for what they once had to put up with.[11] The fact that a decline in private care potential is to be expected in the future does not need to be elaborated here. The most important factors: changes in household and family structures, the emergence of smaller ageing cohorts, mobility, female employment, etc. are well known.

Another fact is the changes that have taken place in the environment of family relationships in recent decades; there has been a change in the roles of men and women. The old pattern of the man going to work and the woman staying at home is no longer correct for a variety of reasons; economic necessities, changed values, etc. play a role. When both partners are gainfully employed, specific stresses and frictions arise in family relationships. The form and functions of the family have, as mentioned above, undergone major changes in recent decades. Solidarity and integration between generations also have to take into account new forms of living together. Furthermore, the burdens that arise for families due to the care of the elderly, unemployment of relatives, disabled children, high mobility requirements for professional reasons, etc., have to be considered more strongly. However, these findings are not concrete indications of a manifest generation conflict.

If we look at the results of studies that focus on intergenerational relationships outside the family, the picture is not dramatic either. At the level of society as a whole, some studies show potential for conflict, but no hostility in the relationship between the generations. Seniors are attributed little involvement in intergenerational conflict. They are still considered to be a group that tends to be disadvantaged, but one that is nevertheless more likely to face restrictions in the future in competition with other disadvantaged groups, such as families with young children. With regard to the question whether the younger active population or the pensioners will have to make sacrifices in order to secure the financing of the pension system in the future, a certain polarisation of opinions is indicated, however not so much in the sense of (age) group-egoistic attitudes, but rather as age-unspecific dissent across the population. Conflicts become more likely where old and young meet as strangers and where clichés and prejudices therefore take the place of personal knowledge. Overall, the relationship between the generations in society as a whole seems to be characterized by a "benevolent ambivalence" (Gerhard Majce) that could also tip into the negative. However, the majority is sceptical about the future of this relationship. Only a few believe in an improvement, half in a continuation of the status quo, and as many as four out of ten assume that the relationship between old and young will deteriorate in the next twenty years.

---

[11] Cf. an early, psycho-socially oriented casuistry in Hartmut Radebold, Hildegard Bechtler and Ingeburg Pina (1981).

3.5   Images of Old Age, Past and Present

3.5.1   Definition of Terms

Do we know about our "spiritual baggage" when we speak of old age? From many conversations I am aware that most people, especially in middle age, know little to say spontaneously on this subject. However, there seems to be one notable exception: when they are confronted with the ageing of their own parents. For after lengthy discussions, an astonishing wealth of initially unconscious knowledge then often emerges. However, this is based less on facts than on traditions, attitudes, assumptions and experiences of rarity. Such "knowledge" can be summarised by the term images of old age.

Images of old age are images of old age that we carry in our heads and that are attached to stereotyped terms. The "elderly in need of help", the "elderly who like to travel", the "old man", the "active old age", "older workers with limited productivity" are key words behind which certain images of old age stand. They determine our perceptions and guide our actions. We communicate with the help of these images and speak about old age under their dictates, and we judge it – our own and that of others. They have very different sources such as culturally and historically dependent models of thought, fairy tales, stories and poems etc. (cf. Radebold 2009). The central question, however, is: what is communicated with the images of old age? In order to get an answer to this, we have to look at the ways in which old age has been talked about at different times, in other words, we have to take a close look at the discourses on old age. But before we can begin this analysis, a number of preconditions need to be clarified.

What is understood here by discourse? Discourses are forms of understanding-oriented communication. They are public or published and thus represent a practice of speaking and thinking. The term discourse does not apply to private discourse, although this is of course never independent of public discourse. It is in this practice of speaking and thinking that the things spoken of are systematically produced. At certain times, the totality of discourses follows certain rules and principles that determine how things can be spoken and thought about at all, what counts as true or false in each case (Amann 2012, pp. 209 and 210). Twenty to thirty years ago, people were still talking about refugees, then, in a creeping process in which legal bases as well as people's consciousness were changed, the asylum seeker narrative emerged from this, which successfully invalidates, for example, the mental link between persecutees from the time of National Socialism and persecutees from today's totalitarian and terrorist countries. The notion of the solidification of

thoughts and ideas through repeated naming in the medium of social groups or reproduction in print media was already an integral part of the social theories of
Wilhelm Jerusalem or Georg Simmel.

The fact that the aforementioned practice of speaking and thinking systematically produces the things that are spoken of falls under the concept of the social
construction of world views. We use ideas and constructions to bring order into the
flood of phenomena. These incorporated mental schemata (quite in the sense of Jan
Piaget) produce relatively solidified ideas with the help of which the present takes
on meaning, other people become understandable, and social space can be opened
up. The constant construction, dissemination and repetition of such ideas of order
sink deeply into our conceptions of the world; they take on the character of social
facts to which we orient our actions. Of course, this immediately raises the question of why some constructions or images of age are so particularly powerful and
can even persist against better knowledge, while others experience a boom and
then fade into the background or disappear completely. These mental constructions
are "social" constructions. They are developed, people discuss them, appropriate
them, and come to the same or similar views about a thing, so that relatively fixed
beliefs emerge. Through the media they are disseminated, repeated and confirmed
(Amann 2004, Chap. 1). In their emergence, they are involved in power processes,
both in the private and the public sphere; even the selection and type of presentation are linked to power potentials and their processing. Thus these constructions
themselves become social facts.

As a thing is seen, so it is. The stronger these convictions are, the more they
immunize themselves against information and knowledge that could shake or re-
fute them. In this situation, such oppositional knowledge is often no longer sought.
A fixed "world view" emerges. Thus images of age are communication concepts,
they have always been and still are ambivalent, they are subject to political and
economic changes, and they take shape as culturally temporarily stable patterns of
interpretation.

3.5.2   Ancient Images of Old Age

The type of discourse in antiquity was primarily one of lamentation: lamentation
about the finiteness of life, lamentation about the past years and their joys. Euripides
(485–406 B.C.) formulated: "I hate sad, killing old age; cast it down into the sea!
Never in mortal houses nor in cities shall it enter." One thing is certain, however:
antiquity did not share the view of an original and simple division of life, namely
that between youth and old age, as often happens today – especially in the general-

izations: "*the* young" and "*the* old". It is true that Aristotle (384–322 B.C.) also knows the complaint, indeed the assembly of all negative stereotypes of old age. The old "are malicious…They also easily suspect evil because of their distrust… They are also petty…They are money-hungry…They are also fearful and dread everything in advance, so that old age prepares the way for cowardice." But, it depends on the context in which old age is placed. With his doctrine of the good middle, he formulates ideal personality traits for the ideal life course. Thus, he recognizes three ages, youth and old age being two extremes, but the proper measure expressing itself in middle age. It is not much different with Plato (427–347 B.C.). For him the ideal is the old man (persons older than 50), as he states in the Νόμοι (Laws), he is concerned with the example and model function of the old. This thought leads him up to the praise of the aged, as enunciated in the introductory passages to the Πολιτεία (The State). But let us not be confused: the answers which the old Κέφαλος (Head – he with the beautiful head) gives are in praise of the properly lived, the philosophical age. The complaints, he says, are not, indeed, unjustified, but it is not the defects that matter in the first place, but the right sense of dealing with them. "But the complaints … have one and the same cause: not old age, O Socrates, but the manner of men's senses." Indeed, complaints about the burden of old age have been part of the common inventory of ideas about life throughout antiquity. "When the body is broken by the mighty blows of age, and the waning strength of the joints rusts, the mind slackens and the tongue and spirit go out of joint" said late still the Roman poet Lucretius. (97–55 BC).

Without giving further examples, it is clear: As a type of discourse we are dealing here with the age lament. But: it is a moral discourse in which the right way of ordering life, the right way of dealing with ageing, is presented as a doctrine. What the factual life situation, today we would say the empirically determined standard of living of the old, looks like, plays no role in this normative discourse.

3.5.3 A View of the Middle Ages

From the high Middle Ages up to the nineteenth century, the theologians now become the experts on old age; at any rate, they are the most important spokesmen of the discourse on old age. The type of discourse becomes one of the consolation of old age. It is the consolation of Christianity. An exemplary consolation book in many respects comes from Daniel Tossanus (1541–1602) in 1599. It is a genuine consolation book, not a duty book, as was Marcus T. Cicero's (106–43 B.C.) "Cato Maior." Daniel Tossanus speaks as an old man to the old, he sees the old man as in need of consolation and as a theologian in a position to grant consolation. Now what about the afflictions of old age in his case? They are there, and they are not to be glossed over, but neither are they to be condemned. It is a matter of raising up the weak, doubting man, and therefore he considers conversion and forgiveness especially important and possible in old age, after a long life. It does not help to lament about the vanished youth. The old should be pious, because that is the condition for facing and enduring all the adversities of life and old age. Now what is the center of this reasoning? Christian piety is the most important power of old age, but it is not attainable without effort and exertion, not without conversion and prayer. And, of course, all the negative aspects of old age are no excuse for lack of piety and zeal for the faith, but if these are present, they become the consolation of old age.

Without again giving further examples, it is clear that as a type of discourse we are dealing here with the rust of old age. Again, it is a moral discourse in which the right way of ordering life, the right way of dealing with ageing, is presented as doctrine. The framework of the teaching, however, is no longer ancient philosophy, but Christianity (which, however, has its roots in antiquity and in the Old Testament; too little consideration is often given to the fact that our thinking is a heritage from Babylon, Athens, Rome and Jerusalem). What the factual life situation, today we would say the empirically determined standard of living of the old looks like, again plays no role in this normative discourse.

3.5.4   The Invention of Old Age as a Social Problem and Economic Burden

Today, it is a fact that older people are the largest group receiving welfare state transfer payments, that pensions are the largest expenditure item in social budgets, and that older people are the most frequent consumers of social and health services. Behind the development that has led to this situation were, in addition to demographic ageing, social strategies and constructions corresponding to them (Amann 2012, p. 211).

In the first thirty years after World War II, old age was identified as a social problem. The first "warnings" interpreting the older as a "burden" on society date from the early 1950s. The United Nations spoke of "the burden of population ageing" and Konrad Adenauer, in the great government declaration of 1953, threatened that it would be the old who would be affected by the decrease in the number of working people, due to the decline in births, in the population. These interpretations, which provided the new phenomenon of population ageing with initial classification markers, occurred in parallel with the consolidation and expansion of national pension systems in what came to be known as the Golden Age of the Welfare State. The main socio-political goals were the efficient removal of older workers from the labour market and the creation of relative income security in old age Amann (2012, p. 209).

This strategy had two effects: On the one hand, labour force participation in the higher age groups (60–65) gradually declined, so the number of pensions increased; on the other hand, it was increasingly accepted that the income needs of older people were lower than those of the "economically active". Thus the so-called income replacement rate is now below 70% in Germany and just above 60% in the UK. In political language, old age, or the beginning of old age, was equated with the statutory retirement age. Old age had become a social problem that had to be regulated under the direction of social policy (Amann 2012, p. 212).

Then the construction of retirement as a solution to economic problems emerged. Starting with the labour market problems, the international economic crisis ("oil shock") and the fiscal tensions in the early/mid 1970s, a changed reconstruction of the social meaning of old age began in two ways: On the one hand, age was redefined from the retirement definition, which had been attached to the statutory retirement age, into a much broader category ranging from 45/50 to death; on the other hand, the extreme increase in the number of early retirees due to social policy early retirement options massively exacerbated the devaluation of older people in the labour market. It was at the beginning of the 1980s in both Germany and Austria that early retirees were denounced in the media, while at the same time the transfer of older workers to so-called soft jobs or even their dismissal increasingly became the most popular displacement strategies in companies (Amann 2012, p. 212).

In the 1970s, this process had begun with a massive drop in male labour force participation in the older age groups in particular (with exceptions in Sweden and Japan). This process was essentially demand-driven by the employment collapse from the mid-1970s to the early 1980s. This socio-politically initiated massive trend towards early retirement meant that there were, on the one hand, those who saw early retirement as a desirable alternative to unemployment and, on the other, those who were effectively driven into early retirement by a hostile and age-discriminatory labour market. During this period, the construct of the lack of economic productivity of older workers spread, equally as a model of legitimacy (Amann 2012, p. 212).

The socio-political strategy in many European countries was to relieve the strained labour markets, i.e. the economic system in general, by mass retirements. In Austria's iron processing industry, as of March 1983, older employees (52+ years for women and 57+ years for men) were forced into early retirement by a decree of the then Ministry of Social Affairs in order to relieve the corresponding sector of the labour market. Significantly for the age constructs, from the moment this "help-the-economy-through-the-pension-system" was recognised as no longer financially viable, those retiring early were redefined as scapegoats for the increasing burden on the pension system. Today we are faced with the paradoxical situation that in the EU in general attempts are being made to raise the employment rates from a low point which the policy itself has brought about, and that at the same time the over 45/50s are being discriminated against in terms of employment opportunities, career opportunities and further training on the labour market. The older had gone from being a social problem to an economic burden (Amann 2012, p. 213).

At the end of the eighties of the twentieth century, the coverage in the West German press on the subject of old age was one about assistance for the elderly. The old people to whom attention was paid belonged primarily to the clientele of social welfare: they were those in need of care and nursing, the poor, the sick, the lonely and those abandoned by society. For years these reports were supplemented, also in Austria, by advertisements, especially in illustrated magazines. Insofar as these were directed at old people, they were almost identical with pharmaceutical advertisements. Symptoms of listlessness, fatigue and dejection characterised the pictorial and linguistic representations. The old person was described with imprecise generalizations. He was portrayed as an old man characterized by manifold wear and tear, marked by manifold deficits and suffering from premature, or in this sense timely, old age complaints. The pharmaceutical industry and the numerous associations and organizations that are committed to the well-being and care of people, especially those in need of help, have played a considerable role in this image of old age (Amann 2004, p. 16).

As a type of discourse we are dealing here with the burden of the elderly. Here it is a political-moral discourse, the factual life situation, i.e. empirically determined standard of living of older people, plays a constant role in this normative discourse. In this way it differs from the historically older ones.

3.5.5   The Euphoria Discourse of Old Age as a Counter-Im age

Now, in a strong countermove, seemingly positive images have been appearing again for some years. Enjoyment, eternal youthfulness and material prosperity are highlighted. In their positive exaggeration they are again one-sided and therefore false. Just as a short time ago the malaise of growing older was the main topic, so now its avoidance, even its prevention – anti-ageing and forever young – is the rage of all prejudices. "I'll get older later" is one of the many book titles that can virtually provide the motto for these images. Quite specifically, these images describe the young, active, mentally mobile, sociable, communicative, healthy, physically fit and sporty, sometimes even politically rebellious old people. This "new age", a discovery of science, is thus characterised by creativity, a wealth of behaviour, independence and autonomy, freedom from the need for outside help, social integration, a variety of interests and a focus on leisure and consumption (Amann 2004, p. 17).

As a type of discourse, we are dealing here with one of old-age euphoria. Here it is again a moral, but at the same time also an ideological discourse, above all now one of youth-centered aesthetics, and the factual life situation, i.e. empirically determined standard of living, again plays a constant role in this normative discourse. Moreover, all discourses on age are obviously also moral discourses.

Our constructions determine the world as we see it and they guide our actions, as do our images of age. At the level of political action, collective constructions have an effect; they result in laws and regulations and shape the world institutionally. Collective constructions are also at work in our everyday actions, and they result in the ways in which we treat our elders. The most important task is to find out which constructions are negative, why they are negative, and who can have an interest in using them.

A Brief Overview of Historical Forms of Participation in Old Age

4

Recent research offers the idea that the structures of early societies on the subsistence level of hunting and gathering were family-based and egalitarian (among men). The structures of social order in Pleistocene societies were the same as those of recent hunting-and-gathering societies with regard to their constitution (Günter Dux). Kinship, family, myth and religion stood in the centre. Only within such structures the situation of the old becomes understandable. Looking back to pre-industrial times, the hypothesis may be: Family economy and work organization were so closely intertwined that a transition from working life to retirement, understood in today's sense, was not structurally foreseen and had no place in the consciousness of those involved.

4.1 Findings from Ethnology and Cultural History

At the beginning of this chapter, a few brief remarks may help to take a look at how age and age groups are constructed in traditional societies. Georg Elwert (1992) assumes that in all non-industrial societies one thing is missing: "an assignment of social status to age counted by years, to chronological age" (Elwert 1992, p. 261). Accordingly, a chronological concept of age is also missing. Therefore, the question arises as to how an age classification comes about in such societies (if it should be perceptible). The answer corresponds to a rough typology with the following characteristics. In savage societies, concepts of age and ideas about the old are based on physical abilities. Further, there is a differentiation that applies especially to women and is based on positions in the reproductive cycle; the decisive factor is whether women are marriageable, married, mothers, divorced, widows, or mothers-in-law. Unmarried and childless old women or men may never become "old". Third, seniority systems are observed in which age is always defined relatively in relation to offspring. Those who grow old without being an "elder" may be marginalized. Finally, age and generational class systems are mentioned, in which groups of peers move up into other age categories in ritually determined annual cycles (Elwert 1992, pp. 261 and 262).

4.1.1 Knowledge and Seniority Principle

The status of older people, i.e. the prestige and status as well as power they enjoy, is dependent on the structures and values in place. Upward and downward tendencies with regard to the status of old age can be effective side by side in history (Rosenmayr 1990; Borscheid 1989; Minois 1987). Only in this way can it be understood how diverse and markedly contradictory the findings present themselves, which are still to be discovered prehistorically, reconstructed historically from traditional societies, or observed ethnographically in retreats of tribal cultures (cf. Lee and Daly 1999; Diamond 2013). One summation that can be drawn in this regard was articulated by Jared Diamond (2013) as follows: In some traditional societies, he argues, people are forced to neglect, abandon, or even kill their elders (see also Amann 1989a, pp. 11–13). However, it is highly probable that such practices are restricted to nomads and societies in particularly inhospitable environments. Other traditional societies provided their elders with a far more satisfying and productive life than most Western societies. Jared Diamond, in agreement with numerous anthropologists, suggests that factors enabling such variation included environmental conditions, the usefulness and power of elders, and the values and rules of the society in question (Diamond 2013, p. 44). They are mostly segmentally differentiated, based on kinship or family systems as well as hunting, production and living communities.

For tribal societies, Leopold Rosenmayr notes that in them, in various modalities, the principle of anciennity or seniority prevailed. It served to maintain stability in the face of cultural change, albeit slow. Information or knowledge, especially sacred and mythical knowledge, "passed in stages from the elders with the higher consecrations to the younger ones, who could advance in knowledge and consecrations if they obeyed the system of rules represented by the ancients" (Rosenmayr 1990, p. 41). When "elder power" is spoken of, it expresses a seniority principle that operates broadly and comprehensively, with inherent control by elders, but which applied more often to men than to women. Experience and accumulated knowledge become a trump card that always stings (power potential). Since these societies are without writing, such knowledge must prove itself in oral transmission and subsequent use. Thus, if old people possess practical knowledge for survival and tradition (e.g. myths of descent, etc.) on the basis of their experience, they inspire respect; they must appear indispensable to the younger ones. At the same time, as old people they are already partly in the realm of the dead, in the realm of the ancestors, on whom, in the imagination of many of these societies, everything depends for the well-being, success and duration of the people. In this perspective, some of the elders gain fearsome power. They play the mediating role between the living and the ancestors (Amann 1989a, p. 46). It can be surmised with some justification: In order to safeguard the memory of creation myths, traditions, magic and rituals, it makes sense to communicate them in words and images to as many as possible, who can then carry them on again. The elders thus transfer their memories to the memories of those who follow. This transmission is likely to be the basis of relations between people and between generations. I suspect that this is related to the traditions of storytelling in so many pre-industrial societies, narratives that at the same time lead to the conception of time as a form of narrative (see also Brock 1998).

4.1.2 Material Factors and Valuation System

But it is not knowledge alone that confers status and prestige. The power of disposal over land, over the people of an extensive and widely ramified clan also plays a role. Clearly, there are marked differences in the importance of each of the factors mentioned for the power of the ancients. Where the settlement is sparse, the power of disposal over land may play a lesser role; where the soil is productive and the climatic conditions advantageous, the need for "many hands" may prove less. Prestige and influence are distributed differently between men and women and – women are not always inferior. Of decisive importance, however, is again and again what manner and level of development production has reached (Amann 1989a, pp. 46 and 47). A distinction, however precise, between hunters, gatherers, cattle breeders and crop farmers already shows clear differences in the conditions and forms of old-age prestige (Simmons 1945).

In connection with it and shaped by it is the value system of a society, from which in turn the reputation and esteem of the old flow. On the Fiji island of Viti Levu, Jared Diamond reports from his own experience, the older people live in the circle of family and lifelong friends in the village in which they have also been at home all their lives. Often they also live in the house of the children who look after them. This goes so far as to "pre-chew and soft-prepare food for an elderly parent whose teeth are bruised to the gums" (Diamond 2013, p. 246). As far as this may be from our practices (and our internalized valuations and assumptions of taste), don't mothers in our country, when spoon feeding of infants begins, sometimes take their food into their mouths beforehand? But values and rules and the actions that follow them are contingent. For the social conditions referred to here, it can only be said, with all due caution, that it is material and immaterial factors because of which older people are regarded as useful, and because of which the younger can support the older. The cultural system regulates respect for the older, respect for privacy, it regulates the preference of the family over the individual (from which generational conflicts can arise) and responsibilities. Again in other words, "Always, then, the level of development of the division of labor and of production, as well as the extent of the forces that the maintenance of subsistence devours" (Amann 1989a, p. 47), will be a decisive framework condition for the cultural shaping of all other expressions of life.

4.1.3   Where Life Is Meager

In fact, there are numerous sources from which materials of ethnological and cultural anthropological field research have been collected in archives and to which authors refer again and again. There are, for example, the Human Relations Area Files, Inc. (HRAF) in New Haven, Connecticut; there is, for example, the Cross Cultural Cumulative Coding Center at the University of Pittsburgh.[1] These archives hold data on thousands of societies in which investigations and observations have taken place. I won't dissipate here about the source-critical and methodological problems of all this material.[2] In any case, this is the background for the claim that there are many reports or cases on which we can rely.

Under extreme poverty there is only the present, the future comes into view only under the aegis of the ancestors. The poverty that oppresses people in a small manageable group with the most primitive means of production, almost independent of other groups in the environment, is of a different kind than that in which millions

---

[1] Just typing the titles in quotation marks into the Internet brings up thousands of entries.

[2] There is a long tradition of critical reflection on social anthropological and ethnological work and its results. This already begins with the well-known researchers such as Lucien Levy-Bruhl, Bronislaw Malinowski, Edward Evans-Pritchard or also Victor Segalen with his idea of exoticism, understood as an aesthetics of the diverse, which in my opinion has been little received (cf. Segalen 1983). Recent criticism now attempts to integrate epistemologies from different disciplines (cf. Hafner-Fink et al. 2017).

find themselves in the world today. The cases in which the elderly are abandoned or killed occur in societies "in which the elderly become a serious burden dangerous to the safety of the group as a whole" (Diamond 2013, p. 250). That this situation is somewhat likely to occur among nomadic hunter-gatherers and, on the other hand, in climatically inhospitable areas (arctic regions and deserts) has already been mentioned, and yet it is not the case that they all sacrifice their elderly. "Some groups (including the !Kung and the African Pygmies) seem to have a greater reluctance to do so than others (for example, the Ache, Sirionó, or Inuit)" (Diamond 2013, p. 251).

From the relevant reports, a kind of gradation of behavior patterns can be read. It starts with neglect until the old people die. They are ignored, given no food and allowed to wander off. So it is reported by the Inuit in the Arctic, the Hopi in the North American deserts, the Witot in tropical South America, and the Australian Aborigines. Another kind of neglect is accompanied by some activity. Of the Lapps (Sámi) of northern Scandinavia, the San of the Kalahari Desert, the Omaha of North America, and the Kutenai and Ache Indians of tropical South America is recorded: They intentionally leave old or sick people behind when the rest of the group moves camp (Diamond 2013, p. 251). Simone de Beauvoir cites a case of the Sirionó in the Bolivian jungle.

Holmberg[3] tells of the eve of a general departure: My attention was drawn to an old woman lying ill in a hammock, too ill to speak. I asked the head of the village what was to be done with her. He referred me to her husband, who told me that they would leave her here to die (…) The next day the whole village left without even saying goodbye to her (…) Three weeks later I found the hammock and the mortal remains of the sick (Beauvoir 1977, pp. 41 and 42).

Among the active inducement is that, for example, the Ache took the men in the woods to a "white man's road" and let them go away. Never again was anything heard of them. Another version of behavior is documented for the Chukchi and the Yakuts in Siberia, the Crow Indians in North America, the Inuit, and the Vikings. The older person is told to commit suicide by jumping off a cliff, going into the sea, or seeking death in battle. An old helmsman from the Reef Islands in the Southwest Pacific said his official goodbyes and then set sail in a boat from where he never returned (Diamond 2013, p. 252). Decisively intrusive appears to be another way that can be described as euthanasia or murder on demand, for example by strangulation, stabbing or burial alive. "Among the Kaulong people of southwestern New Britain, it was common even into the 1950s for a widow to be strangled by her brothers or son immediately after the death of her husband" (Diamond 2013,

---

[3] The anthropologist Allan Holmberg, who published a book about the Sirionó in 1948.

p. 253), in the service of an "obligation." Jane Goodale is mentioned as a warrant person. The last version describes the victim being killed violently without her own cooperation or consent.

For the most part, these accounts come from the sources of ancient ethnology; the reality reported here has changed, has passed, and the authenticity of the accounts is sometimes subject to embarrassing questions. On one thing, however, the research seems to agree: Poverty becomes a determining factor, capable of stifling sentiment but also effectively preventing the development of habits and institutions that would ensure care for the sick of the aged and the "useless." These small communities are so fragile that poverty constantly threatens their existence.

Nevertheless, it should be noted that in most societies older people were not neglected, although "there is nothing natural about providing for the elderly" (Elwert 1992, p. 265). They were asked for advice, in different ways and rituals, they were given food either regularly or from abundance. It can be summarized as a baseline from many studies that while in almost all of these societies the elderly held a special position, their knowledge played an important role and status and prestige were attached to it. Here we can once again refer back to the human being as a "deficient being", for old age can also be seen as an over-imprinting of this specificity, as could be said with reference to Paul Baltes or Arnold Gehlen. This phase of declining physical strength, which is certainly essential especially for hunters, warriors and nomads, this phase of declining mental agility, this phase after menopause is a key period for an institution-building of care to compensate for deficiencies (Elwert 1992, p. 265). In any case, the care of the older may have been particularly dependent on whether a social order had been developed that went beyond individual interests and assigned functions to old age (the stages of old age). The actual treatment of the older, however, always depended to a considerable extent on the mode and state of production, the distribution of goods (itself culturally regulated), the division of labor, and the mythico-religious conceptions and normative arrangements that had developed with these very conditions (Simmons 1945). Anthony P. Glascock (1986) reported on a 1983 study in Somalia (Bay Region) of the control of property and rights among old men. In the transfer of property and disposition rights (in the patriarchal Islamic society studied), daughters were disadvantaged compared to sons and a distinction was made between "good" and "bad" sons. Good are those who are submissive from a young age and indicate that when their fathers will be weak, sick and decrepit, they would take care of them and support them until death. In the selection of the criteria which make a good son, the old men interviewed were strikingly unanimous, and likewise in the fact that only the longest possible insistence on rights of ownership and disposal over land, watering-places, and animals, would ensure provision in old age and the attention and help of children. This explains the economic inequality between young and old men frequently mentioned in the above sources, but in the case of matrilineal tribes such as the Lele in Zaire the situation is different (Douglas 1971).

In traditional societies, a horizon can be drawn for the question of the participation of the old people in life that could not be more diverse. Culturally regulated ways of involvement/disengagement with the concomitant of social integration range from a position of dominance secured by status and prestige, which literally has to vouch for the continuation of the tribe, to presumably general benevolence, to gradually increasing denial of the means of participation, which in the extreme case concludes with the active extinction of life. Arguably, the construction of forms of life in all these ways includes, on the part of individuals, a kind of emphatically learned acquiescence to these arrangements. Effective and repeated revolt against these practices by elders does not seem to be reported anywhere in the sources. Involvement/disengagement are highly regulated in traditional societies: according to ages and initiations, according to gender rules and kinship networks, according to mythico-religious rituals, and according to rules of behavior designed to perpetuate reasonably conflict-free social relations. The attempt to maintain functioning social relations between different groups, for example as reparation after an injury, can lead to extremely complex behavioural rituals and require an expenditure of time and energy which, from a Western perspective, may appear downright abstruse. A compensation of the material damage in an equivalent way is far in the background, centrally it is about reconciliation. In such processes, too, the experiences of elders are called upon and needed, because they know how other cases were resolved in the past (cf. Diamond 2013). Participation in life in archaic societies may be interpreted as highly regulated on the one hand and as structurally undercomplex on the other, since the regulations that are actually effective stem from a structure in which economic production still takes place *in* society (Karl Polanyi) and hierarchies of domination do not yet exist. But even these relations are shaped quite differently depending on the role played by property and rights of disposal, prestige and influence, knowledge and magic.

4.2   Participation in the Age of Industrialisation

To speak about the situation of the old in the age of industrialisation means to be confronted with that historical constellation in which the strengthening capitalism and the state (in almost all countries still in the form of the monarchy) begin to develop the new structures of rule in which the old people begin to crystallise as a group of socio-political interest. Parallel to this, however, there still exist older (class) structures of the pre- and early capitalist period, which now come into conflict with the new developments.[4] From the multitude of constructions developed in historiography, social and economic history and early sociology to describe or explain this change, I pick out the so-called "family economy". The dominance of this form, originating in the domestic community, began to give way to non-domestic wage labor with capitalism's demand for labor. While it was assumed in early social history that the household community did not presuppose a "house" in the modern sense, it certainly presupposed a minimum of planned farming (Weber 1922, p. 195), and then later a system of gender- and age-specific division of labor and roles, as well as of property and domination (Amann 1989a, p. 89). For a long time a view was formative, e.g. represented by Otto Brunner (1956)[5], which nourished patriarchal and autarkism romanticizing ideas at the "house"; in the meantime criticism (e.g. by Claudia Opitz) has corrected essential aspects. For the sake of clarity, I will differentiate between peasants and craftsmen, since their "family economies" differed greatly from one another.

4.2.1 Rural Life

In the sense of an ideal-typical definition of characteristics, the peasant business enterprise formed a relative unity of production, consumption and family relationship system. However, it is precisely the expression family system of relationships that needs to be looked at more closely. Husband, wife and children (the nuclear family in today's understanding), other relatives, sometimes even mentally or physically handicapped brothers and sisters etc. worked on the farm. They were often joined by the so-called servants, day labourers and seasonal workers, depending on the work to be done, the time of harvest, etc. Reinhard Sieder documented (1987, p. 17) that, from a social and domestic law perspective, the far-flung relatives and the servants belonged to the "peasant family" in many areas of Austria, Germany, France and Switzerland. Ideally, therefore, the family was made up of those who belonged to the peasant economy on a permanent basis. Until the

---

[4] Cf. the important "Source Collection on the History of German Social Policy 1867 to 1914", commissioned by the Historical Commission of the Academy of Sciences and Literature – Mainz, from it. e.g. Born et al. (2000a, b); Henning and Tennstedt (2003).

[5] Otto Brunner, an Austrian historian with an affinity for Greater Germany and National Socialism, was the author with whom the concept of the "whole house" as a category of early modern constitution and society took off.

nineteenth century, this family economy produced for its own needs; it was not until the end of the nineteenth century that industrialization led to the replacement of the servants by machines in the family economies, and the peasant family was reduced in its genealogical core to the "bourgeois family model" (Sieder 1987, p. 19).

In these circumstances, a happy and cared-for old age, characterized by inclusion and full integration, as Ernest W. Burgess once drew it in a romanticized manner, was hardly possible (Burgess 1962, p. 350); in any case, there was no welfare state protection against accident, illness, and old age (except in rudimentary pieces of the church poor system). Security and a proper place of existence were found only in the family. Unmarried old people and those left to their own devices, whether men or women, led a miserable life. Peter Borscheid has repeatedly referred to the situation of widows and the inhabitants of nursing homes. This agrarian world provided work, toil and hardship for the majority of its people, precisely the peasant population; they were dependent on nature, at its mercy. Even those who worked long and hard in the fields had no guarantee of a good harvest. A single storm could destroy everything. So it is also to be understood that the thinking and efforts of the peasants revolved around the preservation and increase of land ownership, because survival, well-being and social standing depended on it. Necessarily, this also placed inheritance in a central position, whereby from generation to generation the handling of property was one of the most important basic categories of life (Borscheid 1989, p. 320). For the peasant sector, the most important structural element of integration and at the same time disengagement of the elderly was the "Ausgedinge", which recognizably held some advantages for the peasant economy, but predominantly disadvantages for the old peasents concerned. For all the rudimentary provisions, the "Altenteil" meant for the farmer that he lost his function as head of the farm (although the obligation to cooperate did not usually cease completely, neither for the old farmer nor for the old farmer's wife), but the successor was now obliged to provide for the livelihood of the elderly. Often petty and extremely detailed handover contracts show how much the assurance of rights, such as getting meat for dinner on Sunday or being allowed to use the main entrance of the house, were the result of conflicts (or means of avoiding them). In pre-modern societies, under-valuing of old age, exclusion and discrimination almost certainly also occurred in the rural sphere. The old themselves, in order to undermine the structurally supported disenfranchisement, often tried to hand over the farm as late as possible,[6] or, after the handover, to assert themselves as

---

[6] This pattern is consistent in its logic with the attempt of old men in archaic societies to delay the transfer of property and power of disposal as long as possible.

indispensable by providing work and services to younger people. A study in the early 1980s in Upper Austria showed just such patterns and strategies as still existing today.[7] The old farmers and farmers' wives "move into retirement", but continue to try to prove themselves useful, "what can still be done"; the relationship between old and young is not always easy, "one must somehow keep one's mouth shut too"; at least the young daughters-in-law admit that the help of the old farmer's wife relieves them, "the mother can only sit in the kitchen at the table and prepare the cooking and look after the little one, but that helps me already." Looking back to pre-industrial times, the hypothesis may be: Family economy and work organization were so closely intertwined that a transition from working life to retirement, understood in today's sense, was not structurally foreseen and had no place in the consciousness of those involved. There will, however, have been a regrouping and restructuring of work areas. Work that was very physically demanding, work in remote locations or work that required great skill and was dangerous was probably taken away from the old. Even today, the large, electronically controlled agricultural machines are rarely driven and operated by old farmers; that is the domain of the young. Presumably, however, there has always been a clearly discernible division between men's and women's work; within the Ausgedinge it is still visible today (Amann 1989b). In any case, the position of the older people in rural society has experienced ups and downs, depending on the reputation and status of old age itself, with lows before and until after the Thirty Years' War, and with highs from the middle of the seventeenth century to the beginning of the nineteenth century.

4.2.2   The Situation in the Craft Sector

The situation of the craftsmen was structurally different. Family life and economic life were also closely interwoven among them, but the social form of the whole house did not apply. In most cases there was no household property and the separation between household and business was clearer than in the peasant case. The frequent material confinement and the rules of the scarcity society determined the

---

[7] In 1983/1984 I had carried out a study project with students of folklore and sociology at the University of Vienna as part of a two-semester course. The topic concerned ageing in the rural world, the project included various qualitative methods as working tools, also a longer field stay in a selected rural community in Upper Austria. Due to delays, which are common in such projects, the project report was completed in 1986, but not printed until 1989 (Amann 1989b).

forms of old-age provision and the situation of old people in the crafts quite substantially (Borscheid 1989, p. 359). Nor should it be overlooked under any circumstances that the living conditions in the craft trades depended to a considerable extent on the efficiency of agriculture. Crop failures, state mismanagement, crises and price increases in the peasant economy constantly depressed the crafts. The consciousness of insecurity dominated man the older he became. Constant lack eats ulcerously into body and soul, it stirs up distrust and the craftsmen saw how more and more guild members slipped into a caretaker existence on the fringes of society due to lack of work and underemployment (Borscheid 1989, p. 359). Werner Sombart drew a vivid picture of these conditions.

> Many small masters live with their families in such a restricted and unhealthy manner that they often almost provoke the pity of the inspecting officials; the whole family, large and small, whether healthy or ill, lives in a small room, which is the workshop, apartment and kitchen; here they work, cook, wash, sleep from early in the morning until late in the evening in summer and winter, usually without any other air exchange than the opening of the door brings with it (...). Sometimes the woman has to help in the business herself; this is often the case with tailors, slipper-makers, shoemakers, furriers, cap-makers, bookbinders, etc. But it is above all in the trade with the goods that she has to work. Above all, however, it is up to her to trade in the products produced. (Sombart 1902, II: 563/564 footnotes)

Although the guild rules determined that the master's household included his wife and children, but also the apprentices and journeymen living in the household, the decisive difference to the peasant business lay in the production process (Sieder 1987, p. 103). In the case of the peasants, everything depended on ownership and the availability of cultivable land; in the crafts, the means of production were easier to procure, but the publicly organised moment of qualification, of technical planning competence was added.[8] For the craftsmen, the close ties to land, which for the peasants was the only guarantee of the necessities of life, but also an iron chain for the generations and a thorny burden, were eliminated.

The hierarchy between apprentices, journeymen and masters, organized and legitimized by the guilds, created relationships that were different from those in the peasant business. In addition, masters who possessed great skill and knowledge of craftsmanship may have been held in high esteem, which was probably qualitatively

---

[8] In the literature, it seems to me that there is an undue valuation of handicraft knowledge and skills over peasant competence. If today, for example, we take the tools made by the farmers themselves for alpine dairying, haymaking or timber harvesting in our hands and may marvel at how they have proven themselves practically, for centuries (and even the latest technical aids are often copied from the old forms), then doubts may arise about this assessment.

different from that of large peasants, although this becomes questionable at a time when the onset of capitalism with its development of manufactory and factory increasingly put traditional craftsmanship under pressure (cf. Amann 1989a, p. 92). It may be that the situation was characterized by a dichotomy. On the one hand, some craftsmen were able to command status and influence because of their recognized qualifications, their guild-secured rights vis-à-vis journeymen and apprentices, and because of his paternal authority. On the other hand, as already mentioned, basic economic security was insecure enough. A certain compensation could be provided by the small ownership of land on which agricultural subsistence farming was practiced (Amann 1989a, p. 93). In contrast to the farmer, the young craftsman was now not dependent on inheritance of his means of production; the sons, who as a rule had not done their training in their father's workshop, were not waiting for the inheritance, and so the master craftsman, unlike the farmer, did not have to "hand over" during his lifetime. The sons went on the move, looking for their own workshop, often by marrying a master's widow or daughter (Sieder 1987, p. 110). These characteristics of the craft began to disappear at the latest with the advent of trade regulations and the abolition of trade restrictions.

The processes of involvement and disengagement were organized in a class-like manner among the craftsmen, with often little room for maneuver for the individual; integration took place almost exclusively through the village community among the peasants and through the guild organization among the craftsmen, whereby the density of settlement may have played a certain role among the latter: Craftsmen in the village or in a small town lived differently from those in so-called guild towns with a certain affluence. Old age was not a status of being provided for without complaint; for the peasant, his own children had to provide for his old age, while the craftsmen provided for themselves, including for their surviving dependents through the many widows' and orphans' funds in mercantilism. Work in various forms often lasted until the end of life and women were almost always in a worse position than men. Nevertheless, integration into the small communities is likely to have been relatively good, even in old age.

Part III
Empirical Findings from Recent Research

5.1   Self-Determined Life

As has already been made clear, a concept of participation that is to be appropriate
to the given conditions of contemporary society cannot grasp the determination of
the guiding concepts from a philosophical, supra-temporal standpoint of justification, but must remain committed to these conditions. Self-determination of the
subjects is therefore to be gained from the context of a democratically constituted
society under the conditions of a capitalist market economy. Democratically constituted means that the constituent elements of this form of state or government,
namely equality and freedom, must be considered in their concrete form. This was
already the case with Alexis de Tocqueville and has not changed since (Tocqueville
2016[1835], p. 15 ff.). Now, however, self-determination is an open and iridescent
concept. For the purpose put forward here, we hold that it has found its political
translation in the postulates of freedom and equality (Dux 2013, p. 15), and in a
very specific configuration: freedom calls for self-determination in the shaping of
the practical conduct of life, equality, on the other hand, anchors in the firm ground
that self-determination cannot take place anywhere else than in the form of life
created by the people themselves (Dux 2013, p. 15). This gives democracy a specific purpose. It is the form of a social constitution whose task it is to create the
framework conditions for a self-determined way of life of the subjects in the first
place, because this cannot be expected from the economic system. However, the
present democracy is a form composed by power, and the question must be asked
where it prevents many people from leading a self-determined life precisely because of this. A first indication is of a principle nature: people are prevented from
leading a self-determined life wherever they are subject to organised power, i.e.

© The Author(s), under exclusive license to Springer Fachmedien   99
Wiesbaden GmbH, part of Springer Nature 2023
A. Amann, *Living - Participating - Growing Old*,
https://doi.org/10.1007/978-3-658-39681-7_5

domination, and are therefore dependent and have to shape their lives according to the dictates of this domination. In present-day democracies in the shape of capitalist market economies, this applies at any rate wherever people have to exploit their labour power under the dictates of the ruling system. This is also the great contradiction in all currently visible versions of the concept of participation: In principle, it is only possible in the present society within the framework of the economic system, which at the same time is unable (and unwilling) to create opportunities for participation for all (cf. Amann et al. 2010a; Amann 2008b). This also makes it clear that all social forms and institutional arrangements that create or want to create opportunities for participation outside the economic system are normatively planned task bearers that correspond to the role of gap fillers. It is no coincidence that the participation of older people, and this is what we are concerned with here, accounts for the smallest share in the economic sphere. This assessment is not a pejorative judgement and does not question the integration services of these institutions, but it should make clear that the present society has created these opportunities for participation (mainly via the cultural system) because the economic system, which pretends to be the central system of this society, cannot or does not want to provide such integration services equally for all people.

It is part of the basic human constitution to have to assert oneself in life, and every subject tries to comply with this principle by striving to arrange his life in such a way that he can understand it as a meaningful life. Self-determination arises from meaning. However, it would be an infantile understanding of self-determination if it meant nothing other than that ultimately every human being has to determine for himself what is to be considered meaningful for the way he leads his life (Dux 2013). The possibilities of choosing a way of life and forms of living are predetermined by society, and it is out of this conditionality that it is important that the subject can and does make use of them. I have set this out in the concept of "learned dispositional scopes". Self-determination is obviously a concept that is determined by the social as well as the individual power to shape the conduct of life and always oscillates between coercion and autonomy. There can be no question that the democratic constitution's stipulation of self-determination for the conduct of one's life is measured against economic possibilities. The fact that work and education are the most important categorical (but questionable) forms for a successful way of life that can be subjectively experienced as the meaning of one's own life in contemporary society is evident, but will not be further elaborated here (cf. Dux 2013, p. 60 ff.).

5.2   Societal Areas of Participation

In current reflections on the role of democracies and their internal functionality, there is a consensus that the establishment and consolidation of general and equal participation rights and obligations, as well as their use by the population, are among the indispensable features of democracy, whereby participation aims at inclusiveness and competition at liberalisation, at least this is also true despite strong tendencies to the contrary in some European countries. At the same time, it seems that in recent years the political debate has placed more and more emphasis on the second aspect. At times it seems as if the principle of participatory democracy, which was once the guiding principle for the further development and deepening of the European Union, has slipped into a status of lesser importance. For a long time now, attempts to assess constitutional rights and social participation have focused on institutional aspects. The reason is probably that the assessment or measurement of institutional features is seen as relatively unproblematic. For example, the nongovernmental organization Freedom House produces an annual report, Freedom in the World, in which it assesses the degree of democracy and freedom in countries and disputed territories around the world. The report is based on extensive checklists on the realization of political rights and civil liberties, which are used to make assessments by country experts on scales ranging from 1 (freest) to 7 (least free). Thresholds are then set for classifying a country as "free" (1.0–2.5), "partly free" (3.0–5.0) and "not free" (5.5–7.0). Freedom in the World captures three dimensions in the area of political rights and four dimensions in the area of civil liberties Fig. 5.1.

However, the anchoring of participatory principles in the constitutional system of a country says nothing about the practice of social and political participation. For not all participation rights are used by all citizens and, conversely, not all forms of participation in political and social life are institutionally regulated or possible.

Political rights	Civil liberties
Election process	Freedom of opinion and belief
Political pluralism and participation	Freedom of assembly and association
Functioning of the government system	Rule of Law
	Personal autonomy and individual rights

Fig. 5.1 Dimensions of political rights and civil liberties. (Source: https://de.wikipedia.org/wiki/Demokratiemessung and also Campbell and Barth (2009), among others)

This makes the question of so-called participation research interesting: Who participates in what form and with what results in the shaping of political and social coexistence? What is meant by political and social participation? Research on political and social participation is embedded in a long tradition of research. Alexis de Tocqueville described the United States as a "nation of joiners" and he saw this characteristic as a cause for the strength of American democracy (cf. Dux 2013). He made no distinction between political and social participation. In the twentieth century, this view was apparently initially lost. Research devoted its attention primarily to political participation, especially electoral participation. It is only in the last 40 years that research has returned more strongly to the tradition of Alexis de Tocqueville. The terms "civil society" or "civic society" were used as catch-all terms for people's political and social engagement. Even though these two types of behaviour are closely related, it is nevertheless useful to separate them analytically, because they fulfil different functions and are directed at different addressees. Today, this perspective applies above all to research in political science and, in a narrower sense, in democratic theory. In social gerontology, a somewhat different tradition has emerged.

In it, it corresponds to the research practice to make a division of the social totality into economic, social, political and cultural sectors and to separate these from each other. This kind of differentiation originates in sociological theories of society and has gone so far in the history of this science that even systems corresponding to these labels have been distinguished from one another, each supposedly functioning autocratically according to its own rules. Although many arguments could be used against this, and the social developments of the last 50 years make it increasingly clear that these divisions are becoming more and more difficult and are losing their heuristic force, they are retained here. It would serve no discernible purpose to attempt to completely rename the terms and their relations in the course of a conceptualization of participation. I therefore follow the differentiation into economic, political, social and cultural participation. For the sake of completeness, it should be noted here that international organisations such as the United Nations Economic Commission for Europe also use terms relatively arbitrarily in their policy papers. As an example may serve: UNECE (2010), where cultural participation is merged into social participation without further explanation.

5.2.1 Economic Participation

For the results reported for Austria in this and the next three subchapters, the broadest research projects available for the topic at hand have been used: Age and Future

(2010) and Federal Plan (2016) and Amann et al. (2018). In them, an inventory of the entire relevant social gerontological and ageing policy research literature was presented in 2010, which was followed by the so-called "Federal Plan for Seniors" (adopted in parliament in 2012), a programme catalogue with target formulations and recommendations for action for 14 thematic areas of social policy for the elderly, which was then followed by an evaluation in 2016. In addition, Amann et al. (2018) was taken into account, on which the following compilation is primarily based. Practical examples and measures that are important in the context of social policy for the older are also explicitly presented here.

In the relevant research, economic participation is analysed primarily in terms of participation in the labour market and focuses on the situation of older workers and older people outside the labour market. The legitimation of this intensification is pragmatic and relies on a more or less everyday understanding of the point at which people are (or can be) classified as older workers, while the upper limit is usually seen as the statutory retirement age, which at the same time signifies a regulated termination of participation. In this group, the inner contradiction of full participation of all by way of the economy becomes fully apparent. Normatively it is not up for discussion, it is unanimously demanded (even if for completely different reasons than that of a self-determined way of life), but in fact the greatest obstacles stand in its way. Older workers bear the greatest burden of sub-optimally functioning labour markets (highest proportion of long-term unemployment, lowest chance of reintegration), their labour power is seen as inferior (lack of productivity with relatively high expectations), the economy is resourceful far beyond its own reason in attempts to get rid of them; by some media and the few-thinkers they are classified as lazy about work and obsessed with pensions, others speak of reduced performance, health risks and resistance to further education, the general negative assessment of age already hits the 45-year-olds in this way. As is usually the case, women are also more strongly disadvantaged here.

This negative picture, which corresponds to reality in many areas, should now be differentiated somewhat. On the subject of early retirement from working life, there have been signs of a change in thinking in recent years, specifically aimed at raising the de facto retirement age. However, progress has been slow and, unfortunately, this rethinking is not fed by ideas of further participation of older workers but by expectations of cost savings in the pension system. New and old ideas are to be found in the extension of the working life and thus an expansion of the employment of older people. Here, the discussion of an increase in the statutory pension limits is also repeatedly up for debate. However, there are considerable doubts as to how effective an increase in the age limit alone, combined with financial deductions, can be. There is still a widespread desire to retire early, and both those affected and companies are interested in opportunities for early retirement. This also applies to parts of the trade unions. Moreover, there are still jobs where people cannot grow old at all. This is also connected with the fact that some employee representatives inside and outside the companies see an earlier retirement as an important contribution to the humanisation of the working life of older colleagues. This situation is similar to that in Germany (cf. Heinze et al. 2011, Chap. 6).

In the evaluation report on the Federal Plan for Senior Citizens (2016), the following results were reported by Anton Amann and Roland Loos (in summary). The number of jobseekers in Austria had been continuously increasing for about 4 years. The number of persons registered as unemployed with the Public Employment Service (AMS) Vienna had risen by 19.7% year-on-year to 121,769 in August 2015 compared to 2014. In this context, older people were in a worse position: the number of jobseekers under 25 had risen by 7.4%, those over 50 by 24.2% (AMS Wien News 13.10.2015). For older workers, the unemployment rate at mid-year 2015 had been 16.2%, which was almost 7% higher than the overall rate (Wirtschaftsblatt 24.07.2015). So it had to be stated: A sustainable improvement of the situation for older jobseekers is currently not in sight.

In addition to older jobseekers, the long-term unemployed (people who have been out of work for more than a year) of all ages are particularly hard hit by the development of crises on the labour market. Being hit by such crises is one of the most severe effects in the disemployment process, as loss of income, social exclusion, conflicts in relationships, etc. tend to go hand in hand. At the end of August 2015, there were nearly 148,000 long-term jobseekers, up 27,000 from a year ago. Half of the increase was in Vienna, where there was an increase of 13,000 to almost 64,000 people affected (Kurier 15.09.2015). In the case of long-term job-seekers, it was once again the older people who were the most difficult to reintegrate into work and the world of work, so that one can speak of "double exclusion" or even multiple exclusion (taking into account consequential effects) in their case. In any case, discrimination is particularly visible in these groups. In October 2015, 391,000 people were looking for a job in Austria. People with a history of flight (refugees) are not included in this figure because they are not included in the AMS statistics. Only recognised asylum seekers are taken into account by the Public Employment Service. Currently, it was said at the time, according to the AMS, 4.8% (19,000) of job seekers are asylum seekers, two-thirds of them in Vienna. The AMS expects another 35,000 asylum seekers to be looking for jobs in 2016, which is likely to exacerbate the situation on the labour market (Kurier 02.10.2015, p. 7). Due to the refugee movement, both employment and unemployment figures among migrants are likely to increase more strongly in the coming years than previously assumed. The Austrian Institute of Economic Research (Wifo) assumes that a total of 764,000 immigrants will be employed in Austria by 2020. This is about 150,000 more than before. According to the Wifo forecast, the average unemployment rate in Austria will rise from 9.2% to 9.9% by 2017, but starting from this high value, it will decrease again somewhat from 2018 (9.4% for 2020) (Standard 14.10.2015, p. 17). It can be assumed that the rather high number of people with refugee experience, who are also getting older all the time, will bring further deterioration for older jobseekers in the labour market in general (Amann and Loos 2016, Chap. 3). In the meantime, some statistical values have changed compared to the period of the stocktaking, but there is no doubt that there will be no remarkable improvements in the fundamental situation of older workers in the future either, at least not as long as politicians continue to persist in their indecision. However, if there were to be a decisive shift in policy priorities in favour of older workers, the following points would need to be considered.

Contradictions and Worsening Prospects[1]

- the argument of higher average personnel costs, which, however, are justified in different ways, e.g. by the seniority principle or (allegedly) frequently longer absences from work
- the so-called shorter 'half-life' of knowledge in certain sectors, which requires 'up-to-date' formal and specialised qualifications
- an (allegedly) greater need for security among older workers in terms of job content, workplace and housing, which is incompatible with growing occupational denormalisation and demands for flexibility and mobility
- the widespread negative self-perception of older workers, especially among the less qualified, with regard to their own internal and external employment prospects or inter-firm mobility opportunities
- increasing importance of mental workload factors (Heinze et al. 2011, pp. 91 and 92).

Conditions for Better Employment Opportunities in the Future

- Decrease in physical workload factors
- Increase in knowledge-intensive work, which is (supposedly) compatible with age-typically higher levels of experience, overview and contextual knowledge

---

[1] In the following lists, the word "allegedly" in parentheses always means that the empirical research findings on the topic in question are inconclusive and that more intensive research would therefore be useful.

- increasing orientation of products and services towards an overall ageing customer base (which makes the externalisation of ageing in companies appear counterproductive)
- increasing supply of part-time jobs to match the part-time aspirations of older workers
- increasing propensity of older women to work, although this varies greatly from region to region and from sector to sector
- Increase in networked, self-organised and decentralised work, which is (supposedly) compatible with a typically higher level of willingness to take responsibility and knowledge of context and experience.
- (allegedly) more favourable initial health conditions for older cohorts moving up in age
- an increase in higher educational qualifications, which could mean an increased willingness to participate in further and continuing training
- (allegedly) growing insight among older people into their *shared responsibility* for their own ability to work (Heinze et al. 2011, pp. 90 and 91).

Most of the research is devoted to questions concerning the employed, little is known about economic participation through continued work after retirement (even "moonlighting"), through economic activity of the self-employed beyond the legal age limit, such as artists, etc.

In addition to the employment of older workers, the analysis of economic participation in research primarily includes the consumption of older people. It should be noted that a distinction is usually made between (a) participation through employment up to the legal retirement age and (b) consumption by all those of retirement age. We are therefore dealing with very heterogeneous groups from the outset, for whom the idea of participation cannot have a uniform meaning. Moreover, the discussion, at least in the media and in everyday life, is very much permeated by one-sided views. On the one hand, a euphemistic tone prevails which has little to do with the real differentiation in old age (not all older people go on world trips or cruises), on the other hand, there is sweeping talk about the older generation without taking into account different access to markets, purchasing power, etc. The older generation is not the only one. In any case, it would make more sense to consider the respective consumer decisions and consumption behaviour against the background of the position in the life cycle and the specific circumstances in the biography (cf. Sect. 7.2.2). Considerable differentiations would result simply from the respective position in the occupational, family and income cycle, which in turn helps to shape life situations. As in all areas of participation, the multi-dimensionality of action also applies here. After the children (the last child) have left home, older people may feel compelled to change their home, to change their mobility behaviour, or the like. In any case, retirement will lead to changes in consumer behaviour, due to the elimination of work-related expenses, reorganisation and redevelopment of leisure behaviour and social engagement. For some, this will be combined with a shift in activities such as travel, increased spending on wellness and health-promoting activities. In any case, for many older people who are parents or grandparents, this status leads to increased spending in the interest of and for their offspring; this pattern is empirically stable. Noticeable changes in consumer behaviour result from health-related changes such as functional limitations, need for help, reduced mobility (about which the studies on activities of daily life provide masses of information), loss of life partners, increased need for security, etc. Once again, considerable changes result from moving into a different form of housing. If determinants are to be named in detail that have an impact on consumer behaviour in old age, then the following would have to be taken into account: level of education, gender, health (also as a condition for satisfying other needs), residential area, desire for security, desire for independence and self-determination, desire for quality of life, convenience and functional facilitation in everyday life, social contacts, etc. (cf. also: Heinze et al. 2011, Chap. 5).

Products and services that are primarily used by older people have found their way into the collective term "senior citizens economy". However, this does not refer to an independent, clearly definable economic sector, but rather to a so-called cross-sectional market. The products and services of the senior citizens economy are characterised by the fact that they integrate different sectors, e.g. digital technology and housing offers in Ambient Assisted Living. However, this cross-sectional logic also has a conceptual disadvantage, because literally all sectors, from retail to leisure and tourism to financial services and housing, are linked to it. Now, in recent years, it is precisely demographic change that leads us to expect that employment effects and economic growth will result from the development or supply of these partly integrated services. Demand and consumption structures as well as employment effects have been researched extensively. Future topics will certainly be technology and digitalisation, technology acceptance, technology and health. The economisation of old age, which is implicitly and explicitly connected with the senior citizens economy, requires a more extensive critical discussion. In general, it should be noted that economic participation has always suffered from conceptualization difficulties, as it is located in two differently structured areas of society: On the one hand, in the labour market; on the other hand, in the consumer markets, and there with considerable importance in services financed on the one hand with money from private households (leisure, travel, etc.), and on the other hand with money from the public sector (care and nursing), whereby, in the case of pension recipients, the financing of expenditure must in turn rely to a considerable extent on transfer payments. From a sociological perspective, funds, wherever they may come from, are easier to classify as "means to …" than as a medium of participation in their own right. Thus the care allowance (in Austria) is regarded by many recipients, objectively wrongly but subjectively justified, as a component of income, especially by poorer persons. This leads to the mostly frowned upon or at least critically assessed behaviour of using the care allowance to "buy" the gratitude or affection of grandchildren through gifts of money, instead of spending it on the purchase of care or support services (which was intended by the legislator). Finally, on the subject of money as a means to…, reference must be made to the large markets of tourism, leisure activities, the sportswear industry, etc., through which culture, the foreign, the adventurous, etc. can be experienced (consumed) and are vigorously supplied by senior citizens' associations of the most diverse kind.

5.2.2   Political Participation

Politics and political institutions are not unaffected by demographic changes. This is true for elections and the electorate, parties, trade unions and other organisations. Although the age trend in political organizations is largely shaped by the increasing number of older people in the population, demographic processes are not the sole cause of this development. Older people, beyond their demographic presence, are more attached to institutionalized forms of political participation than younger segments of the population, as evidenced by their higher voter turnout, as well as their representation in political parties and trade unions. Prevailing trends in various countries include: The number of eligible voters is growing; eligible voters are getting older; about one in three voters is older than 60; those aged 60–69 are most likely to vote; older people are more likely to vote conservatively. People of older age are under-represented in national parliaments, and there are few women among older MPs. Parties are shrinking, over 50% of members in traditional parties are over 60, the ageing of parties is a long-term trend. Older people are also organised in trade unions, but trade unions are a male domain (cf. also Menning 2009).

For Austria, reference should be made at this point to the "Federal Plan for Senior Citizens"[2] adopted in 2012, in which political participation was deliberately

---

[2] www.bmask.gv.at/site/Soziales/Seniorinnen_und_Senioren/Teilhabe_aelterer_Menschen/

placed in the broader framework of social participation.[3] It contains both goals and recommendations on this topic under point 3.1 Social and political participation.

5.2.3   Social Participation

A general assumption is that older people are socially integrated into society in a variety of ways. Research results and everyday experience prove that they are actively and passively present in social networks, are to a large extent members of families and friendship circles, belong to and participate in associations, do voluntary work, etc. However, this does not change the fact that people are at risk of exclusion as they age. Health impairments, lack of mobility, low levels of education, poor infrastructure, lack of access to services and, finally, ageism are among the well-known barriers to social inclusion. There are a large number of proposals in the social sciences on how integration could be thought of; functionalist sociology even spoke of an *integration function,* by which was meant that the task given to every society of distributing its resources and gratifications was then solved when this distribution was accepted by the majority of the members of society. In this way, however, the issue of integration is seriously misunderstood, because the point of reference of success or failure is shifted back to the subjective evaluation of the individuals alone, which apparently completely ignores the consideration of diverse constitutive environments. In the *microsystem,* it is the processes of socialisation and enculturation as well as identification in the context of acting individuals and their social environment with which they are directly or indirectly in contact; the microsystem can be separated into a *mesosystem* and an *exosystem*; in the first, developments take place in which people are directly involved, in the second, in which they may never enter, events take place that influence what happens in the environment, in both it is a matter of placement in various formal and informal social groupings (social intercourse circles Georg Simmel called them); in the *macrosystem* it is the anonymous mechanisms of institutions such as the social division of labor, markets, money, and expert systems, which determine the integration of people behind their backs, so to speak. Although these topological territories (following Kurt Lewin) deviate from the usual level distinction between micro, meso and macro levels, they have the advantage of being based on action-

---

[3] Cf.: https://www.sozialministerium.at/cms/site/attachments/9/7/1/CH3434/CMS1451919586.368/ soziale-themen_seniorinnen_bundesseniorinnenplan_gesellschaftliche-und-politische-partizipation. pdfwww.bmask.gv.at/site/Soziales/Seniorinnen_und_Senioren/Teilhabe_aelterer_Menschen/   The citation references with "SeniorInnenplan" refer to the document given here at www.

related interrelationships and therefore seem to be far more suitable for analysing questions of the development of social relations in an integration process. After all, it must be taken into account that in the process of integration as many social groups as possible, in this case the older, are brought into the political, social, cultural and economic structures of a society so that they can participate in the decision-making processes on issues that affect them. This presupposes that there is a consensus that exclusion should be minimised and eliminated, and that all those who are disadvantaged should be helped by society, which, according to what has been said above, falls within the remit of an age-sensitive social policy.

Social inclusion can also be understood as a development that involves the building and relative stability of values, relationships and institutions in a society where everyone, regardless of gender, age, origin or religion, can fully exercise their rights and responsibilities on an equal basis with others. Every person should be able to grow old in safety and with dignity, and be able to contribute to society in a meaningful way (see UNECE 2010). Integration and participation are therefore closely linked to the concept of social cohesion, a very important element of an inclusive society. It refers to the ability of a society to ensure the well-being of its members, minimize inequalities and avoid polarization and conflict, and it requires the cultivation and promotion of solidarity and reciprocity between generations (UNECE 2010). This last idea addresses the notion of familial and societal intergenerational relationships and intergenerational relations, which play an essential role in research on social participation – also in the guiding idea of intergenerational justice in the sense of social participatory justice.

In the evaluation report on the Federal Plan for Senior Citizens (2016), Ines Findenig presented the following results, among others, in Chap. 1, which are summarised here. The fact that participation and involvement of older people in society is meaningful and necessary is hardly disputed. The deeper reason for the demand for participation lies in the discrepancy between an inadequate socialisation model of age and the actual living situation of the old, a discrepancy that Hans Thomae noticed decades ago. Participation research would therefore also have the task of working on drafts of social models appropriate to old age. In Austria, there are considerable deficits in this respect. Participation has different meanings, participation is stronger and closer to a continuous or regular activity. Thus, participation is moving closer to *active ageing*.

This guiding principle culminated in 2012 in that it was designated the European Year of *Active Ageing and Solidarity between Generations* by the UN. In Austria, too, *active ageing* has been receiving ever-increasing attention in recent years. This is often justified and reinforced by the hope and appeal of being able to use older persons in various areas beyond the gainful employment phase and, at the same time, thereby prevent tendencies towards loneliness and fragility. Parallel to the danger of instrumentalisation, the opportunity is seen for older persons to also make use of the active ageing image by serving as a positive reinforcement of lifestyle. As already mentioned in the 2010 Federal Plan respect for persons who do not conform to the productivity norm of *active ageing,* this, as a culture of acceptance of non-participation, seems to be only at the beginning of establishment. An up-to-date view of *active ageing* seems to be conducive to subjective life satisfaction in the sense of thinking along. Especially because good (mental and physical) health contributes to increased participation and is at the same time a prerequisite for it.

This connection can also be seen in the fact that especially in the late stages of life, voluntary engagement brings with it a wide spectrum of positive aspects. This ranges from health promotion, stress compensation, increased self-esteem, alternative leisure activities and the expansion of networks and resources to participation processes per se. The dual function of volunteering should also be mentioned here, as it is both a help option for recipients and a clear integration task for the person involved (cf. Heimgartner and Findenig 2017, p. 187). Although there is a lack of concrete comparative data for Austria and the EU, various research studies nevertheless indicate that voluntary engagement or voluntary activity – in the right context – makes people happier and healthier and also helps them to live longer.

Little has changed in terms of gender compared to the 2010 results. According to studies, men over the age of 65 are more likely to be involved in formal activities, which include committee work and political activities. Women over 65, on the other hand, are more likely to be found in the informal sphere. The age group 60+ is further subdivided by gender in the different formal and informal fields of activity. Furthermore, the affinity to (non-)family caring tasks emerges among women over 60 in voluntary engagement. This confirms long-standing, socially traditional and prevailing gender inequalities in volunteering. The higher the level of education completed, the higher the chance of volunteering. Accordingly, incentive systems seem to be necessary that can tend to eliminate such an unequal distribution. Or, in an even better case, it would be necessary to recognise these educational diversities in Austria and to fight for a general improvement of the educational situation, which would at the same time reduce poverty tendencies and thus give voluntary engagement a greater chance. Other trends are also stable: in addition to the level of education and gender, the prevailing image of age influences both social and political participation of older persons.

In 2010, the report Age and Future still referred to the tendency of Austrian seniors to be passively represented by organisations to a much greater extent than to become involved themselves. However, the emergence of a self-confident and *ac-*

tive generation, as outlined in the report, seems to indicate a change. A passive *letting oneself be represented* has seen a partial change in recent years, or rather this passivity is more a phenomenon of those currently over 79. Similar developments can also be observed for informal engagement, which takes place, for example, in the neighbourhood. As new sub-areas sprout up in which participation takes place, other aspects also reinforce the need for new participation formats for people in later stages of life.

Social and political participation is undergoing a further change as a result of discernible trends towards change for older people, not only in the sub-areas but also in the way in which they are active. Framework conditions and structures are undergoing a visible change from lifelong membership in formal associations to project-related, flexible activities. This development is called, among other things, transformation from old to new commitment and is characterized by higher flexibility. Furthermore, this trend also seems to meet the growing self-development motivations in the social engagement of older people. Participation thus seems to be more dynamic and occasion-related than in the past. In recent years, new professionalisation tendencies and quality development processes have taken place in the social participation sector in Austria. These can be seen, among other things, in the implementation of the first Volunteer Act of 2012, through which the Volunteer Council, the web portal freiwilligenweb.at and the voluntary social year are anchored. The law has brought about a small but general increase in the recognition of volunteering in Austria. In addition to the presence of the web portal, which provides structures, information, presentation and networking opportunities for those seeking and looking for commitment, there are now 12 volunteer centres and a large number of volunteer and senior citizen exchanges. Likewise, volunteer fairs have become established in parts of Austria. Volunteer centres, fairs and exchanges represent an important opportunity to sensitise and motivate people to get involved and to organise and support them. At the same time, the introduction of volunteer coordinators or volunteer managers and specific training courses (which tends to be Austria-wide, but is not yet comprehensive) is a future-oriented structuring trend of social participation – especially among older people.

Since the expertise on the Federal Plan for Senior Citizens 2010, a large number of measures have been taken in Austrian organisations to achieve the goals of the plan. In the area of ensuring equal participation (politically, socially, economically and culturally) of older persons, it can be seen that, among other things, individual legal measures have been taken, information channels have been created (e.g. flyers, brochures, conferences, meetings, etc.), projects have been carried out (e.g. in the education sector, in health care, etc.) or other measures (e.g. meetings of the Federal Seniors' Advisory Council, funding calls, etc.) have been initiated or are planned in the medium term. In the anchoring of participation and demands for participation as an aspect of political culture among older people, contrasting tendencies are apparent in the sense of perceived stagnation on the one hand and actions and projects of a positive nature on the other. In the area of voluntary commitment in the sense of taking on social tasks and responsibilities, it can be shown that although a disagreement can be perceived in the comparison of the organisations, a large number of actions have nevertheless been set. In terms of research, however, no outstanding structuring of the survey of social participation (in old age) in Austria can be identified so far.

The awareness of an organized Austria-wide volunteer system in the course of dedicated trainings and curricula for the quality assurance of the coordination of participation seems to be at a hopeful beginning. Nevertheless, a tendency towards instrumentalisation (see *active ageing*) and compensation of full-time volunteers should not be disregarded.

5.2.4 Cultural Participation

Cultural participation as a concept of orientation presupposes that there is such a thing as cultural life and that this is linked to educational processes (Kuhlmann and Naegele 2016, p. 45), whereby a specific facet of education in general, namely cultural education and the path to its participation, comes into play here. Cultural participation then means participation in the artistic happenings (in the broadest sense) of a society via educational processes. Since education is legally secured and institutionally anchored (at least for a large part of society – see the topic of young people after flight), cultural education can be understood as a right, and because existing education produces effects, also as a prerequisite for quality of life; it is a constitutive component of general education.[4]

In terms of empirical participation research, it is useful to distinguish between culture in the narrow sense and in the broader sense. Culture in the narrower sense refers to art/the arts and their productions: Visual arts, literature in its various genres, the performing arts (from theatre to dance to film), music, the applied arts such as design and architecture, as well as the manifold forms of combination between them. They represent the subset of culture in the broader sense that is usually the focus of participation research. Culture in the broader sense is the life processes and practices in their entirety: e.g. from the technical achievements to the behav-

---

[4] Some of the following considerations are based on http://www.bpb.de/gesellschaft/bildung/kulturelle-bildung/59910/was-ist-kulturelle-bildung?p=all

ioural patterns of living together and the values and norms, i.e. also the philosophical and religious reference and interpretation systems of a society. Even more so than in the other areas of participation, a life-course approach suggests itself in the case of culture, because enabling participation in the course of life is closely related to the multidimensional concept of standard of living (e.g. use of scope for action in the course of life). On the societal level, the realisation of participation rights (and obligations) presupposes the provision and design of corresponding participation opportunities and structures, i.e. external resources as mentioned above, and on the individual side internal resources. But what is it about the educational process in this context?

Education is understood as an always newly acquired and constantly changing state or process in which people are able to shape their lives in a self-responsible (and successful) manner. This concerns the use of internal and external resources. Acquired are factual knowledge, practical action competences, emotional competences and the ability of self-reflection, i.e. what is usually called orientation knowledge. Here the terminology in participation research is inconsistent. Insofar as people, their life situations and their reference systems change in the course of their lives, education – to be understood as an educational process – is also never complete. This is the reason why it is often argued that education and learning are a task or duty that accompanies the whole of life, but also an opportunity. This has been expressed very clearly in recent years in the topic of lifelong learning. Cultural education (it is also often referred to as musical or musical-cultural or aesthetic education) means the learning and appropriation process in the ontogenesis of man in relation to himself and his environment in the medium of culture. Ultimately, cultural education means the ability to successfully participate in culture-related communication through knowledge, thinking and action.

Cultural education enjoys considerable esteem in the public discussion, even if the "boom" that prevailed 10 years ago has obviously weakened, hopeful of educational efforts as a whole it still is, whereby a certain degree of social and general education is a prerequisite for the acquisition of and access to cultural education and thus not tangible for every person or life situation. In the context of cultural education, however, the discussion increasingly revolves around the so-called key competencies that young people in particular would be expected to acquire. The cut of the hoped-for successes ties the idea of competence to the economy. Creativity is the key competence most in demand, at least in qualification contexts in the world of work. While it is difficult to teach in pedagogy and didactics from school to vocational training and further education, it is taken for granted as a basic competence in the arts and among many cultural workers, i.e. artists and cultural mediators. The shift of the conceptual framework to the economy inevitably leads to the fact that in the political discussion cultural education is usually appreciated less in its basic significance for the personality development and social participation of the culturally educated, emancipated individual, than for the assumed "soft skills" that are suitable for the labour market and also for integration achievements in the multi- and intercultural situation.

In the context of the standard of living concept, it was stated that these also provide the social forms of organisation in which people can then use their scope and thus shape their situation. Social forms of organisation play an essential role in the question of cultural participation. Like all educational processes, cultural education takes place formally and informally in designated institutions and outside, in the public sphere and at the private level. The general education school system with its subjects of art, music and, where available, performing arts (theatre), plus German and foreign languages, in their literary and cultural-historical components, is the institution in which basically all children and young people, but also adults (biographical aspect), experience artistic-cultural education. The importance of educational processes in the pre-school elementary sector (kindergartens and day-care centres) is also becoming increasingly apparent. Sociocultural institutions and other cultural associations offer opportunities for the cultural participation of older people. This has always been the case for the adult education centres in their cultural departments. The fact that all these offerings are primarily used by the formally better educated is a stable empirical pattern. The professional cultural institutions themselves, such as theatres, orchestras, museums, libraries, art associations, cultural centres, always have a de facto culturally educational effect on their visitors through their work. The projects funded by the BMASK, which have been researched in the present overall project, clearly show that the large area of amateur culture, theatre groups, museums run on a voluntary basis, libraries, art associations etc. plays a strong, often underestimated role in practical cultural education for those active as well as for their audience. The mass media, audiovisual media and print media (from books to daily newspapers), also have a de facto culturally formative effect with their content, i.e. they educate, although to a lesser extent among older people than among younger ones. The diverse possibilities of the Internet contain almost as many opportunities for cultural education and participation, whether in active engagement or in mere consumption.

The fact that cultural participation is primarily achieved through education has been emphasised on various occasions. In the evaluation report on the Federal Plan for Senior Citizens (2016), Franz Kolland and Vera Gallistl came to the following conclusions, again in summary.

In general, Austria shows sufficient documentation of the participation of older people in continuing education and barriers to continuing education based on

European surveys that have been conducted nationally in Austria. According to AES data, participation in formal and non-formal education decreases steadily with age, while at the same time the overall participation of older people in education has increased in recent years. Compared to the 2006/2007 survey data, the data show a slight increase in the 45–54 age group (+5.2%) and a larger increase in the 55–64 age group (+10%). The structural indicator Lifelong Learning describes the share of the population aged 25–64 participating in education and training measures. According to this structural indicator, 5.3% of older people over the age of 60 participated in continuing education in 2015. Participation in education also decreases here with increasing age.

However, this result does not mean that age can be used as an explanatory factor. For it is not primarily age, but other factors (e.g. educational status) that causally explain the decline. Barriers to education for older people emerge that are based less on age than on place of residence, gender and state of health. The school education of older people is a central differentiation criterion in access to education, which influences participation opportunities even more strongly than monthly household income. While education in working life is still male-dominated, women are much more active in post-professional education.

In the last two decades there has been a slow expansion and multiplication of educational offers for older people in Austria. Senior education is located in different organisations and is not limited to adult education offers. This has made it necessary since 2010 to deal more with strategies of quality assurance in senior education. Those projects that have been identified throughout Austria as particularly innovative and creative and that also meet the underlying quality criteria are awarded a prize by the BMASK at the end of the project. On the other hand, in recent years there have been numerous initiatives to disseminate basic geragogical knowledge to people who are active in educational work with older people. On the level of scientific investigation of quality criteria of senior education, the studies "Geragogisches Grundwissen. Untersuchung zur Qualitätssicherung für Bildung in der nachberuflichen Phase" and "Qualitätssichernde Maßnahmen in der erwachsenenpädagogischen Bildungsarbeit in Österreich unter Berücksichtigung der nachberuflichen Phase" should be mentioned.

In addition to these quality assurance measures initiated at the federal level, numerous further training measures for senior educators take place in or in cooperation with a wide variety of organisations. These include the annual workshops in Strobl organised in cooperation with the BMASK and the Federal Institute for Adult Education (BIFEB), training courses for volunteers (e.g. Salzburg Education Network, Province of Vorarlberg) and multiplier training courses (Province of Vorarlberg). Newer developments are emerging in the training of age-sensitive counsellors, who were trained in the form of scientifically based short courses "Educational counselling for active ageing" by the Styrian Education Network and the Salzburg Educational Counselling Service.

In the field of providing age-sensitive information and guidance on educational opportunities for older people, (pilot) projects have been carried out throughout Austria since 2010, ranging from the production of information brochures and flyers to information events and the testing of age-sensitive educational guidance. In addition, two research projects funded by the Federal Ministry of Labour, Social Affairs and Consumer Protection were carried out in the period from 2013 to 2016, dealing with the scientific foundation of possibilities for age-sensitive educational guidance.

In addition to these research projects, (pilot) projects on educational guidance for older people with a focus on the post-professional phase have also been increasingly carried out since 2010. In various federal states, the first activities in the direction of developing educational guidance for older people are becoming apparent. In line with the focus on informal learning as a new, everyday form of learning for older people and the importance of an educational infrastructure for older people close to home, an intensified discussion around the term community education has been established in recent years. The focus is on building an infrastructure that enables formal, non-formal and informal learning, especially for educationally disadvantaged groups, and thus reaches new target groups for further education. The focus on community education was defined in Action Line 6 of the Lifelong Learning Strategy. In recent years, pilot projects have been carried out to build up an educational infrastructure close to people's homes, the innovative potential of which lay in cooperation with regional partner organisations. For example, seven regional conferences of the province of Styria were held in which librarians were trained in dealing with older people under the motto "Reading knows no age(s)". However, no evaluations or further research activities can be found on this subject.

Projects on intergenerational learning have been tested and promoted throughout Austria in recent years. These ranged from projects of older people together with children and young people to the work of visitors of day centres for asylum seekers and homeless people to model projects on intergenerational learning. A practical view of such projects can be found in the guide to intergenerational projects initiated by the BMASK. In general, it is important to distinguish between learning with, learning from and learning about each other in the field of intergenerational projects and learning processes (Franz 2009). Although intergenerational learning takes place in a large number of projects and organisations, it does not (yet) have an institutional framework that could sustainably secure intergenerational teaching and learning arrangements. In this context, it is also evident that scientific research or evaluation of intergenerational learning forms and projects is still insufficient in Austria, with a few exceptions (see Findenig 2017, among others).

With regard to "ageing and media", the "Federal Plan for Senior Citizens" (2012) aims to create nationwide access for older women and men to the new media as well as information on safe use and strengthening their media competence. This is to be ensured by expanding low-threshold, barrier-free and educational offers for women and men in the post-professional phase of life close to their homes throughout Austria. The data show increases in internet use by older people during the observation period, with this increase being higher in the 70+ age group than in the 60–69 age group. For the year 2014, the data are available on a gender-specific basis and show a strong digital divide by gender.

A central challenge of senior citizens' education is the socially unequal use of information and communication technologies (ICT). This is also illustrated by the fact that internet use was a key predictor of the realisation of educational interest in concrete educational participation in a representative survey of people aged 55–75 in Austria in 2013. In recent years, there has been an increase in both research and educational projects throughout Austria aimed at facilitating older people's access to information and communication technologies. On the one hand, the focus is on the implementation of computer courses for older people, on the other hand on the development of age-sensitive didactics with regard to new technologies and the creation of information brochures and guidebooks.

Initiatives include brochures for seniors on the topic of "Using the Internet safely", which was published in an updated edition in 2017, as well as the folder for seniors "Fraud on the Internet – how to protect yourself" (new edition 2017) and the "A1 brochure for seniors: Internet einfach erklärt" (2015), which are published on the saferinternet.at and digitalsenioren.at platform created by the Austrian Institute for Applied Telecommunications (ÖIAT). Since 2010, extensive documentation for trainers working with older people has also been available there. On behalf of the BMASGK, ÖIAT also organised the forum "Seniors in the digital world" for the fifth time in 2018 and published a guideline on "Quality criteria for senior-friendly teaching and learning with digital technologies" in the same year. Also on behalf of the Federal Ministry of Labour, Social Affairs and Consumer Protection, the Austrian Institute for Applied Telecommunications, as the coordinating body, has been offering education providers in the field of "seniors and digital media" the opportunity to have their offerings distinguished as a good practice project for the first time since 2017.

In 2012 and 2013, the Federal Chancellery funded the project "Seniorkom.at – We network the generation". As part of this initiative, intergenerational training courses and events were held throughout Austria, with a special focus on the topic of "e-government". In addition, the production of information brochures on the topic was promoted.

Eva Flicker and Nina Formanek also reported on the issue of cultural participation via media in the Federal Plan (2016) (my summary). The Federal Plan for Seniors 2010 dealt with two central problem areas: media representations of ageing and the *digital divide* between the generations. Media images reproduce prejudices about the old and move between the *idealisation of youthfulness* and the *age deficit model*. The availability of new, digital media and their competent use play an increasingly important role in everyday life today. Quantitatively, it is often proven that older people have less media competence than younger people and they also have less media equipment. At the same time, it must be critically noted that not all older people want to deal with new media and can cope well without them in their everyday lives. Nevertheless, research has shown for a long time that low-threshold and target group-specific advice and information services on the use of new media are an important motor for promoting the media competence of older people.

Since 2010, around 20 new publications have appeared in the German-speaking world on media use or the portrayal of older people in the media. The empirical interest in the field not only remains unbroken, but is becoming relevant for a wider circle of researchers.

Recent studies on the portrayal of ageing in the media can reveal a discreet differentiation. Particularly in cinema films, but also on television, the roles and associated images of old age are more differentiated, closer to everyday life and more credibly staged than in the years and decades before. The thesis that media representations of ageing are one-sided and polarising is refuted. Film analyses are devoted to the previously taboo subjects of dying and death in film. On the whole, however, older people are usually in the minority as protagonists in films. When older actors and actresses take on leading roles in films on the subject of dying and death, it is usually their age that is attributed a particular advertising and impact potential.

In recent years, successful cinema films have broken with the taboos of dying, death and sexuality in old age. However, it remains to be seen whether these more realistic depictions of living conditions in old age will initiate a general trend and increasingly diffuse from the cinema into other media formats. In contrast to the cinema film, which offers new images of old age, studies on other media often show that stereotypical images of old age are still being perpetuated. Particularly in advertising, the cliché image of the young-at-heart, active older people is constantly maintained by addressing them as a target group with purchasing power.

An analysis of leading Austrian and German print media in 2012 shows how even an image of old age that appears positive at first glance, the *young old,* has a discriminatory effect: on the one hand, this functions as a stereotype and contradicts the actual diversity of lifestyles in old age; on the other hand, in conjunction with neoliberal market and advertising strategies, this exerts considerable pressure on older people to remain economically active and gainfully employed, or at least able to buy, for as long as possible.

In the period from January 2012 to June 2015, six agencies of the direct or indirect public administration at the provincial or federal level or interest groups stated that they had taken measures to promote more realistic images of old age. For example, the Seniors' Association has been awarding the media prize, the *Seniors' Rose* or the *Seniors' Nettle* for particularly positive or outdated media images of older people for several years. However, the Seniors' Association notes that prejudiced images of ageing still prevail and that there is still a need for clarification.

Data show, differentiated by age groups, that despite the increasing equipment of households with a PC, currently older people still use the computer significantly less often than younger people, although this may change significantly in the next 15 years. Older users are divided into two age groups here in comparison to the accessible data of the Media Server Study. According to this, the age group of the 65–74 year olds is particularly noticeable; they used the computer in the last 12 months of the survey with 49% significantly less frequently than all other age groups. Similarly, the proportion of people in this age group who have never used a computer is 43%, far higher than in the other age groups.

As in 2010, a gender gap in computer and internet use is evident particularly in the older age groups – most pronounced among those aged 55 and over: 22% of men have never used computers and 25% of men have never used the internet whereas 40% of women have never used computers and 46% of women have never used the internet. These data point to a *digital divide* by age and gender. Mobile internet use with smartphones, tablets, netbooks, etc. has become much more important in recent years. This comparatively young segment of media use in particular makes it clear that younger and older users differ greatly in their use of new media and communication technologies.

Further scientific research into the field is being driven forward in particular by three BMASK studies, on the media competence of women 60plus, on measures for seniors in the digital world and on the practice of senior-friendly product design of smartphones, tablets & co. The BMASK also explicitly offers a project for older women: "Academy and learning network for senior women – tailor-made educational offer for older women in the post-professional phase of life on the use of PCs and the Internet".

The broadband offensive, a legal measure of the Federal Ministry of Transport, Information and Technology, is not aimed at the target group of older people, but the improved Internet access nevertheless benefits older users in particular, as many of them are more dependent on using the Internet at home. The extremely comprehensive study conducted by the Austrian Institute for Applied Telecommunications on "Measures for Senior Citizens in the Digital World" shows that older users express an increased need for security with regard to the use of the Internet, as they are concerned about data misuse and intrusions into their digital and non-media privacy. At the same time, the proportion of online shoppers is growing rapidly in the older age groups. Social networks, e-government and mobile devices or apps, on the other hand, are used less frequently by older users than by younger ones.

Despite a fundamental openness, there are various hurdles that limit the Internet use of older people (e.g. low self-assessment of competencies, lack of resources, lack of usability of devices). Consequently, an appreciative approach to the target group is necessary, which is based on positive images of age instead of prejudices. Furthermore, neutral sales advice is needed in addition to local offers. Under framework conditions for media education offers, numerous practical tips for the design of courses (small groups, flexibility of design, relaxed setting, gender aspects, room and device equipment, etc.) are formulated and detailed practical guidelines on methodology and didactics are recommended.

5.2.5 Contextual Hypotheses on Participation

I call context hypotheses empirically supported conjectures in which individual contexts can be considered in argumentative condensation and emerge entirely from existing empirical results in Austria (in the sense of Otto Neurath). They are taken verbatim from Amann et al. (2018).

Participation in General

In opening up scope and shaping the involvement of people in social contexts, there is one factor which, according to all experience, plays a central role and has a clear broadband effect: education. The level of education has a lasting influence on voluntary engagement among older people, the area in which someone has completed the highest level of education influences the direction in which engagement takes effect. In this context, older people who were not born in Austria show a lower propensity to volunteer than people born in Austria, while the situation is likely to be reversed for younger people; level of education and involvement in citizens' initiatives are closely related, women are disproportionately represented in charitable institutions, while in local history associations or citizens' associations it is again men who participate significantly more often, and involvement in this type of association takes place predominantly in smaller communities, here a structural moment of community building in the small-scale area has a lasting effect in the direction of inclusion and integration across the generations; the income and wealth of a household have a positive effect, higher levels of urbanisation have a negative effect on the extent of volunteering and the province is in itself a differentiating factor; from retirement age onwards women engage less than between 35 and 60 years of age, with the number of people in the household participation in volunteering decreases among older people, among middle-aged people the reverse is true. The chance of volunteering in a managerial role is much lower for women than for men, on the other hand, the influence of gender is not so strong for executive and administrative roles; moreover, the level of education has a stronger influence on volunteering in a managerial role than on volunteering in an administrative or executive role; other volunteers in one's own household are more important for volunteering in an administrative or executive role than for volunteering in a managerial role. Different types of groups and associations prove to be a suitable framework for different people to become actively involved; women are often found in religious groups and in the charitable sector, men in home clubs, sports clubs or in the voluntary fire brigade, people living in a partnership are more often active in citizens' initiatives, while widowed people are more likely to find a field of activity in senior-specific institutions, etc.; here, traditional role models, habit formation, ingrained recruitment patterns and ideational gratification systems play an essential role. The vast majority of people who are involved in groups or associations do so for up to 10 h a month, and these are often people with lower incomes. Otherwise, however, higher income tends to go hand in hand with more intensive involvement. Similarly, people with higher incomes are more often involved in several groups or associations at the same time. Younger people (51–65 year-olds) are particularly active in associations that primarily serve individual needs (self-help groups, hobby clubs, collecting clubs, sports clubs and social associations), and the aforementioned influence of higher education is also reflected here. Employed persons are conspicuously often active in associations that are community-oriented (charitable organisations, voluntary fire brigades), as are persons from larger households. A mixture between these two types is formed by people who are involved in group-specific associations (citizens' initiatives, business or professional associations, local history and citizens' associations): They too often do this alongside gainful employment and live in larger households, but they also often have higher educational qualifications. Social participation of older people in the sense of successful involvement is quite obviously dependent on both acquired and ascribed characteristics and can only be ascertained by taking external and internal resources into account. The repeatedly demonstrated patterns of gender-specific participation and inclusion are a clear expression of the effectiveness of learned dispositional scopes.

Economic Participation
In general, financial deprivation increases with higher population density, but decreases with higher educational attainment; manifest poverty (risk of poverty and financial deprivation occur together and have the effect of strongly decreasing integration) also occurs much more frequently in densely populated areas; more highly educated or socioeconomically better off persons have a larger social network due to higher social capital and, accordingly, more non-related network partners than persons with less education or lower socioeconomic status. lower socioeconomic status; Especially women (of all age groups) and single households, respectively widows and divorced persons are confronted with a low income; furthermore, apart from the connection with formal school education, a significant connection with the health situation and mobility is shown: poverty correlates with health problems and lower mobility, respectively with social withdrawal; there is also a correlation between poverty and social withdrawal. Poverty correlates with health problems and lower mobility or with social withdrawal; there is also

a correlation between employment status and the use of training opportunities, this also applies to the level of income, the state of health, the level of education, age and gender; for employment, age and health are two of the most important influencing factors, age and employment correlate negatively, health in turn influences employment positively, healthier employees are generally employed longer. People with a lower socio-economic status and thus lower incomes or smaller pensions are more often found in the group of those with little contact with their children. Difficult life situations seem to be directly involved in the reduction of involvement.

Cultural Participation
Younger age, higher schooling and better economic situation lead to higher educational activity; similar effects are produced by a large kinship network and employment (which in turn is naturally related to younger age), with long-term patterns being important: the more regularly vocational further education has taken place, and the more regularly private further education has been practised, the more likely there is to be educational participation in old age; it is influenced by school and work experience and goes hand in hand with gender-specific role attributions and with the socio-spatial living situation, the following social groups can be assessed as disadvantaged in old-age education and thus more deprived than others: High-aged people, retired people, people working in the household and people from small localities (thus the most important factors that have an integration-rejecting effect have already been mentioned), if there is a biographical anchoring of educational participation, then organized learning is more likely to occur in old age as well. The cultural offers in the neighbourhood are mainly used by the younger old people, but older people who are active in associations also use these offers more often, those who only have elementary and secondary school qualifications show a clearly lower degree of use, as do those with low incomes; training mental abilities through educational activities seems to be among the strongest motives, followed by deepening knowledge and meeting others, for the so-called educationally deprived strata, costs, recognition and training of mental abilities are more important; not to be underestimated is the subjectively perceived learning effect that emerges when

people engage in numerous activities, although there is a clear difference between men and women – overall, among women the proportion of informal learners and those distant from learning is higher, which takes on particular significance under the fact that participation in education leads not only to greater social integration but also to expanded social participation.

5.3   Objectives

1. Ensuring equal political, social, economic and cultural participation of older women and men
2. Anchoring participation and the right to participation of older women and men as part of the political culture
3. Increasing the participation of older women and men in volunteering and in taking on social tasks and responsibilities.

The *recommendations* stated:

1. Upgrading the political participation of senior citizens' associations.
2. Consideration of further target groups, in particular enabling comprehensive participation opportunities for older women and men with special needs.
3. Clarification of the performance potential of older people in society, motivation of older people for social commitment and voluntary/unsaleried work, and ensuring structures for voluntary/unsaleried work.
4. Ensure comprehensive documentation on participatory culture in Austria.

Within this comprehensive report, the topic of political and social participation has numerous cross-connections with other areas such as: Economic situation, social differentiation and intergenerational justice, older workers and "work" in old age, education and lifelong learning, age and gender issues and the special situation of older women, intergenerational relations and intergenerational relationships, housing conditions and technology as well as mobility, social security, social and consumer protection, securing infrastructure.

Contents of the cross-links are, for example: "self-determination, empowerment and dignity of older people in all areas of the economy, politics and culture (improve) in order to promote their involvement" (Senior Plan: 13). "Among the differentiations among the elderly that have been repeatedly demonstrated empiri-cally are, in the sense of relatively stable patterns of inequality: the disengagement or lack of involvement of certain groups of the elderly and among the elderly" (Senior Plan: 14). "Relatively poorer social situations are often associated with disengagement, care problems, reduced participation, disability and risk of needing care" (Senior Plan: 14). "Not all social cohesion initiatives can be addressed through cash flows. Fair distribution of resources, combating prejudice or discrim-ination of any kind against older people and supporting socially inclusive activities and social networks are among them" (Senior Plan: 14). "When discussing disen-gagement, the following aspects should be considered: [...] social exclusion or lack of integration and participation" (Senior Plan: 14). "Work in old age encom-passes a wide range of activities that are beneficial in multiple ways, both for the worker him/herself and for his/her social environment (such as volunteering or caring for relatives in need of care). This differentiated view should also be taken into account in the discussion of the relationship between working life, retirement strategies and demographic ageing. In any case, social involvement, employment policies and social security policies are closely intertwined from the point of view of work" (Senior Plan: 15). "Education is the factor that plays the decisive role in almost all areas of life, from health to social engagement and involvement, to qual-ity of life and activity interest" (Senior Plan: 20). "It tends to be the younger old, persons with higher levels of schooling and higher incomes, residents of larger places of residence, and persons who are socially integrated who attend classes and organized educational events. Older people who continue their education are more likely to volunteer, have more confidence in political institutions, and are more politically active" (Senior Plan: 20). "In addition, participation in continuing edu-cation leads to social integration or reinforces a positive social image of old age, enhances physical and psychological well-being, increases anticipation and pro-cessing of critical life events, and has a positive impact on civic engagement or volunteering. Education in old age contributes to social participation" (Senior Plan: 21). "In many political and social spheres, opportunities for older women to have a say have not been realized; this is evident in the involvement of older women in political processes of decision-making at national, regional and local levels" (Se-nior Plan: 22). "In general, it must be noted that women's contribution to the shap-ing of society is less visible and also often inferior to that of men. This is a funda-mental contradiction, which must be balanced out or even resolved as one of the most urgent political tasks" (Senior Plan: 23). "For years, the demand for a cross-cutting intergenerational policy has been raised again and again. It would have to help shape generational relations. Empirically, the lack of systematic attention to the equal value and equal status of people of different ages in all decision-making processes is particularly visible" (Senior Plan: 24). "Conversely, communal living makes high demands, not only in terms of suitable living space, but also with regard to the social skills of the residents. The prerequisite is a communal attitude that goes far beyond that of a non-committal neighbourhood. A frequent basic problem in many projects is that older people are often primarily interested in housing, but less in the community" (Senior Plan: 27). "Since social exclusion is an almost inevitable consequence of such minimum standards, the central task for the future lies in the compensatory improvement of social security. The more older people are involved as consumers in diverse markets and become the address of advertising strategies, the more their protection and the adaptation of offers to their needs is necessary" (Senior Plan: 31 f.). "Increased well-being and greater social participation, even in an age-related limited social environment, would support a meaningful lifestyle into old age" (Senior Plan: 43).

Part IV

Versatile Interpretations

Literary Involvement and Disengagement

"I confess that I need stories to understand the world," Siegfried Lenz once said, and this makes sense to me, because our everyday life consists of stories. In what follows, I do not take a sociological approach to literature, as would be suggested, for example, by the work of Pierre Bourdieu, systems theory, or British cultural studies. Why, then, should a sociological analysis legitimately engage with literary materials and hope to draw insights from them that might not accrue to it from recourse to the material it itself produces, that is, its "data material" in the broadest sense? The possible and certainly contingent answers to this question could be based on the discourses associated with the above-mentioned approaches, but it seems to me that not much more would be gained from this than a position of constantly self-perpetuating relativizations. Last but not least, one would also have to contend with the peculiar fact that the sociology of literature in the German-speaking world has never managed to become an institutionally broadly secured and theoretically well-founded field of research from which methodological approaches and theoretical constructions could impose themselves, even though interest has clearly increased since the 1990s. The book by Helmut Kuzmics and Gerald Mozetič (Kuzmics and Mozetič 2003), which is still very much worth reading today, provided sufficient insight into this more than 15 years ago. Also, a focus solely on empirical analysis of conditions of production and reception of literary works, as dominated sociology of literature for a while and thrived on the glamour of the empirically sophisticated, could contribute little to the question at hand. This kind of result showed that the knowledge and wisdom of those who carried out these analyses was not far off by then. Less harshly, one could say with Karl Kraus: "But it is now the fatality of appearances, whose matter is splendor, that to the observer in the moment, and as through this himself, the seen appearance transforms

© The Author(s), under exclusive license to Springer Fachmedien Wiesbaden GmbH, part of Springer Nature 2023
A. Amann, *Living - Participating - Growing Old*,
https://doi.org/10.1007/978-3-658-39681-7_6

itself into the recognized appearance" (Kraus 1917, Heft 445, p. 134). However, inspiration alone, which emanates from literature, cannot be enough either.[1] Against this background, the question posed at the beginning can hardly be answered.

The path I follow takes two well-known literary works of art as examples for a participation-theoretical interpretation, in the same way as was done with the ethnological material and the empirical-sociological in earlier chapters. To do this, some delimitation considerations are necessary. The literary representation of social and cultural constellations is of both aesthetic and sociological interest. The sociological consideration of literature is concerned with its own representation of socio-cultural meaning in literary texts, whether they are "highly" valued, such as Johann W. v. Goethe's "Hermann and Dorothea", Thomas Mann's "The Magic Mountain" or Stefan George's poems, or "lowly", such as Peter Rosegger's songs sung by the lumberjacks in the forest. In this sense, literature can be viewed with profit "from the outside," without therefore having to deny the intrinsic right and intrinsic value of the work of art as a concrete totality. However, since the autonomy of the work of art/text is a socially conditioned autonomy, the task of critique inevitably accompanies this observation, because otherwise it would only be a stare at a fiction of a reality that lies ahead of it, because literature does not describe reality one-to-one, it fakes the described. Thomas Bernhard experimented with this idea: "What writers write/is not against reality/(…)/but everything they write/is nothing against reality/reality is so bad/that it cannot be described/no writer has yet described reality as it really is/that is the terrible thing" (Heldenplatz, p. 115). At the very least, critique here means contradicting convention-bound norms and actions, seeing the fragility and dissonance of the world, and resisting false harmonizations. At the same time, we find a commonality with scientific observation. Art and sociology undoubtedly have a common point of departure: this is "the immediate context of meaning and being (…) in which man is involved with all his dispositions and inclinations, interests and aspirations, his whole thinking and willing" (Hauser 1973, p. 7). Arnold Hauser understands this life context as a totality that is preserved in the attention of art, while life loses this character in scientific access. Science abstracts from the concrete, heterogeneous and atomized material of life, whereas the following applies: "Of all forms of consciousness, art is the only one that resists every sensualizing abstraction from the outset and persistently" (Hauser 1973, p. 8). Art wants its object to become an immediate vision, science goes for the abstract, but what they have in common is "to find out what the world we are

---

[1] Cf. the interview with Sina Farzin: https://soziopolis.de/verstehen/was-tut-die-wissenschaft/artikel/wie-geht-es-eigentlich-der-literatursoziologie/

dealing with is like and how we can best deal with it" (Hauser 1988, p. 5); what they have in common is that they have an epistemological function and are bent on practical goals. In the background of this view, however, looms the old dispute, as it were between Immanuel Kant and Georg W. F. Hegel, as to whether the "truth" of art is intangible or even an arbitrary moment, and whether the view that art and science belong to the same sphere of being becomes a question of metaphysics.

I will not pursue this problem any further here, because in the long look back, even the views of Georg Lukács or Theodor W. Adorno, who once took vehemently opposing positions on this issue and by whom Arnold Hauser, like many others, was influenced, have had to bow to the relativization of time; I consider another perspective to be more significant for sociological analysis. It lies in the idea that all possible and conceivable expressions of life by human beings are anchored in their everyday world, take their origin from there, and have an effect on it. The theoretical constructions and methodological positions corresponding to this principle, which have gained weight in the course of time, beginning with Henri Bergson, Edmund Husserl and Max Weber, continuing via Alfred Schütz and Thomas Luckmann, up to the more recent positions of everyday sociology, must be assumed here to be known, at least in outline.[2] Now, the interweaving of cognitive intention and practical goals has to be put into words, and it needs not only a social, but also an inner-psychic anchoring. A premise of the everyday sociology just addressed is that people live in a world of meaningful relations; we do not confront pure facts, but always experience them in their meaningfulness, in their meaning for people. In other words: Our world is always already interpreted, everything about it is invested with meaning, and because meaning is directed toward something, it can be assumed that our experience is already determined at its source by our human purposes. Alfred Adler derived from this starting point what he called the "meaning of life" and said of it, "But no human being can live without meaning. We always experience reality through the meaning we give it; not in itself, but as something already signified. It is therefore natural to suppose that this sense is always more or less imperfect, unfinished, even that it is never completely correct. Our world of meaningful relations is a world full of errors" (Adler 1979, p. 13). This is an old idea that came to flower with the Renaissance at the latest, and at its core is meant to denote our relationship to the world: Whatever people do and in whatever form it takes, they do it in order to recognize a reality that is chaotic for them in and of itself, to judge it more correctly, and to cope with it more successfully, and indeed to shape it. However, it should not be forgotten that in this process people (social subjects, as Leo Kofler called them) (have to) submit to a context of

---

[2] For an excerpted but condensed account, see Amann (1996).

meaning that is always already given, which has its origin in the rationality, the inner logic and the respective provisional design of the social process. This may also be seen as an initial condition for participation, which on the one hand is tied back to the phylogenesis of the species Sapiens, but on the other hand must be newly acquired and practiced by every human being in ontogenesis, whereby he is dependent on an environment in which he submits to these contexts of meaning or deals with them. That our world of meaningful relations is a world full of errors necessitates another important insight: just as most theories of the work of art assume that the genesis of a work is not finished when the author or creator hands it over, so too a scientific insight, once found, is not final. The process of ongoing, complementary and changing interpretation, of constant reinterpretation, applies to both.

The following two narratives are taken, in accordance with the previously mentioned preconditions that are to justify a sociological analysis of literature, as a form of expression of contexts of meaning of the everyday world and placed in an interpretive context, the interpretation of which is to support the decision as to what the epistemic yield for participation is.

6.1   "The Community Child" (Marie v. Ebner-Eschenbach)

Austria's defeat at the Battle of Solferino on 24 June 1859 and the federalist constitutional law of 20 October 1860 are commonly associated with the beginning of a liberal era in Austria and the unification of Italy (cf. Zeyringer and Gollner 2012, p. 293). In Moravia, where Marie v. Ebner-Eschenbach was born on 13 September 1830, late feudal conditions may have begun to crumble at that time; industrialization had already taken hold with factory work, and the young teacher, a newcomer to the village in the novel, refused the traditional inaugural visit to the landlady at the castle. The village child, the main character Pavel Holub, therefore finds himself in a social world characterised by the village with its notables and poor, the sugar factory with its workers, the castle with its landlady and servants, the church and monastery with their authorities and subordinates, and the countryside with its peasants, day labourers and smallholders. Marie v. Ebner-Eschenbach places the events in her narrative in the middle of this time: "In October 1860, the final hearing in the trial of the brickmaker Martin Holub and his wife Barbara Holub began in the provincial capital of B." (Ebner-Eschenbach 1985, p. 5). Martin Holub, the father of Pavel and his sister Milada, a notorious drunkard and family tyrant, had killed and robbed the priest in the sacristy, was hanged, and the mother, accused of being a confidant, was sentenced to 10 years in prison. As a child and a boy Pavel had known only work, hunger and beatings, was devoted only to his sister and was conspicuous for his obdurate silence.

After the parents were no longer there, the children fell to the community, which did not want to take on this burden. "They do not have any relatives who could be obliged to care for them, and no one will do so out of love" (Ebner-Eschenbach 1985, p. 10). The mayor succeeded in placing Milada with the lady of the castle, who then gave the girl to a convent, which was considered a respectable educational institution, where she was to die before being clothed as a nun, the "loss of his only happiness," said Pavel; but he was given to the parish shepherd Virgil, a rogue, who, together with the cottagers with whom he lived, was one of the most disreputable in the village. No one wanted to see the son of a robber and murderer as a housemate of their own offspring. An even harder time began for the boy, evil gossip pursued him, every prank in the village, every theft, every misdemeanour was laid to his charge, the stupidity and baseness of the villagers seemed hardly surpassable (…) "and he was filled with a boundless contempt for stupidity, which believed the most nonsensical thing about him, if it was only something bad. He found a pleasure in upsetting anew at every opportunity the stupid people who were ill-disposed towards him" (Ebner-Eschenbach 1985, p. 31).

The family, which had immigrated only shortly before the murder, was not yet integrated into the social world, when the father already began to prepare Pavel's future role as an outsider by drinking and refusing to work, by beating his wife and children. Manslaughter and execution, as well as the incarceration of his mother, finally threw him offside. No one wanted to have anything to do with the son of a murderer and a convict. The lady of the castle had a distant dislike for him, the priest, representative of that authority which claimed to judge good and evil, wrong and right, was driven by the same prejudice and vileness as the crassest village idiot. Around the age of 14 a secret, never openly admitted love began to stir in Pavel for the beautiful but frivolous and cunning daughter of the parish shepherd in charge of him, Vinska, that is, his foster-sister, which was only to bring him harm. When the ragged and raggedly dressed boy is given a pair of new boots by the old teacher, his only patron, "proper boots with high shafts," she steals them from him at night, inciting him to all sorts of foolish and questionable deeds, including, finally, plucking out the last great tail feather from the old peacock of the lady of the castle. When the enraged peacock lashes out at Pavel and pecks him on the head with his beak, he strangles him. Now the lady of the castle is finally hostile to him. The people lamented that he would end up on the gallows like his father, stones flew, words worse than stones hit him. "Pavel gazed jauntily about, and the consciousness of ineradicable hatred against his fellow-men feasted and steeled his heart" (Ebner-Eschenbach 1985, p. 49).

The objective external conditions of his life situation could not be more dismal. A childhood that takes place on the shady side from the very beginning can hardly be shattered any worse. His participation in his fellow social world takes place almost exclusively through negatively evaluated actions and negative attributions by other people, the scopes in which he can and must act are scarce and volatile, prejudice, malice and stupidity stand around him like a wall, his exclusion is institutionalized. Since he is also named as a ringleader in thefts from children, one might suspect at least a temporary involvement in a subculture. Nowhere, not even in the family into which he has been placed, can he find a spark of loyalty or support; here we have the not exactly frequent case of participation in social life taking place in a highly selective manner and entirely in a negative connotation. Even sporadic school attendance and work in the factory do not change this. In addition, at that time social integration for children was probably associated with a generally high degree of coercion anyway; the pedagogy of being a free child had not yet been invented.

But, what is Pavel's reaction, how does he "process" these experiences? That he became obdurate, silent and withdrawn into himself has already been mentioned. He tends to remain silent in the face of false suspicions, even when vindication might yield success. "I am silent when I am right, because's pleases me when people are so stupid and then I can think well: You asses!" (Ebner-Eschenbach 1985, p. 62). This attitude condenses into something even more damaging to himself; he wants others to perceive him as a deviant, and he loudly labels himself as such. When the teacher who was well-disposed towards him asked him desperately, "What is to become of you?", Pavel stretched, "put his hands in his sides and said, 'A thief'" (Ebner-Eschenbach 1985, p. 30). After much trouble, he manages to visit his sister in the convent, who is now to become a novice; she tells him that she is the most well-behaved among the many good children, and then exclaims in a tone of conviction, "You are also good?" "Me?" said Pavel, "how can I be good?" (Ebner-Eschenbach 1985, p. 59). This disturbs the sister and she asks him whether he is doing anything wrong; this makes him uncomfortable and he asks in return why he should not do anything wrong, "there is no other way." This passage in the novel is one of the most depressing, because it throws a very bright light on Pavel's inner condition, on his hardships and on the defiance born of these hardships.

"And what wrong are you doing, for instance?"

"Like what? … I take things away from people …".

"What kind of things?"

"How you ask? – What should I take? What I've always taken. Fruit or turnips or wood…".

Then the little girl, in rising fear, yet still hoping, cried out, "Then you are a thief!".

"I'm one too." (…) "How can I not be bad? The parents have been bad, too."

"Just because!" (…) "Don't you understand? – Just for this reason I am the most good in the whole convent and *you* must be the most good in the whole village … so that the good God may forgive the parents, so that their souls may be redeemed … Think of the father's soul, where it is now…" (Ebner-Eschenbach 1985, p. 60).

The sister had already learned her lesson in the convent, but Pavel still knew nothing to draw from his experience of inferiority other than demonstrative defiance. A person who sees himself placed in a hostile world always thinks only of himself, of his need, of what he lacks. And in his defiance he can feel confirmed, because he wants to be accepted as a worker in the monastery's agriculture in order to be able to escape the forced context of stealing, cheating and being cheated in the village. But he addresses his request in vain to the matron, who is a close friend of the baroness at the castle. For she had forgotten the people for the love of God. "And as she gazed before her, infinitely pious, infinitely impassive, so did her whole retinue, and the hard-to-comprehend Pavel understood at last that all his pleading was in vain" (Ebner-Eschenbach 1985, p. 68).

And yet, in the conversation with his sister, Pavel is open to a completely different idea that Milada gives him when they talk about what he is doing for his mother, who is still in prison. He should buy a field and build a house on it, so that his mother will have a place to live after her release. The childish sister gives her savings to the angry Pavel, who has no money for such a project, but can warm up to the idea. "Milada triumphantly brandished a knitted bag, through whose wide meshes it flashed bright and silver" (Ebner-Eschenbach 1985, p. 63). Thirty-four guilders. You don't buy a field or build a house with that, but you can do something, which is what Pavel did then. When he was back in the village, he showed off his treasure in a quarrel over a copper coin in the inn, whereupon the people immediately accused the rag, the beggar of theft, got physical, and he got into new calamities. "Scarcely escaping, the pursuers at his heels, he still shouted back. 'Where did I get's? – Stole I's'" (Ebner-Eschenbach 1985, p. 76). Pursued by the mob, he rescued himself to the old teacher with the speaking name Habrecht (he who is right), who protected him, calmed the rabid pursuers at the front door and showed him a hiding place for his bag in the floor. Intercourse with this insightful person seemed to quicken Pavel's reflections. He had wanted to stay at the monastery, they had turned him down, now he wanted to start at the teacher's. "What" (…), "what to begin?" "The new life" (Ebner-Eschenbach 1985, p. 80).

In all seriousness, a new life, but what was it to look like? He wanted to cultivate the field belonging to the school, but neglected by the community, live with the teacher, board and lodging, work in the factory in late autumn and winter, for a daily wage of one guilder, save and buy a field and build a house. The teacher did not quite want to believe in it, especially since the consent of the community would be necessary. As it turned out, the community leaders, above all the very ill mayor, were unable or unwilling to bring themselves to give their consent. Through dark channels, involving Vinska once again, a herbal potion from the hands of Vinska's mother, a "quack", reached the mayor, who died immediately after Pavel brought it to him. Now he was also accused of murder, initiated by Peter, the mayor's son, who hated Pavel because he wanted the Vinska for himself. The suspicions increased terribly, Pavel had to fear for his life until the gendarme pulled him out and took him to the district court. The mayor's autopsy in the city and the trial before the judge proved that the man died of natural causes and Pavel was innocent. He had spent 2 months in detention, during which time he had crafted a dainty model of the house he planned to build. The 2 months had changed him; he had become a very strong young man, who worked steadily, day after day, for the next year, in the sawmill, in the sugar factory, in the forest, saving the money. Then he came of age, and with the consent of the community and strong endorsement by Anton the blacksmith, who was later to become his friend, bought a sand pit from their landed property at a high price. The community child had acquired property, but the girls and boys were still lying in wait for him and shouted: "Poisoner…! Bist doch ein Giftmischer" (Ebner-Eschenbach 1985, p. 106). In the spring he began to cut the bricks for his house himself, but during the night unknown persons kept destroying his work. Peter married Vinska, Vinska's mother died, Pavel was not moved by all this, he took it with indifference. The priest summoned him and said: "You have been wronged" (Ebner-Eschenbach 1985, p. 11), but when Pavel complained that the children were shouting poison at him and that the adults had uprooted the small spruce trunks he had planted weeks ago, and when he asked the priest to put a stop to it, this specialist of the right faith knew no better answer than that Pavel himself was to blame for his bad reputation and that he needed people's faith in him for his welfare here on earth. But the priest dealt Pavel the deepest blow by telling him that the teacher was leaving. After 21 years of teaching, he was transferred from this village, far away, and he left without saying goodbye to Pavel, who was deeply offended. His best friend had left him without a word.

Pavel finished building his house, spent some time in the military, now finally grew up, a lad of "athlete's strength", saved the life of Peter, his worst enemy, by pulling down a fence post on which Peter was in danger of being crushed by a newly bought harvester which he had set in motion in arrogant recklessness, was accused by Vinska, who had seen nothing of the incident, that he was to blame for Peter's injuries. The landlord, who owned the fence, persuaded the municipality that Pavel should pay for it, and also told the perplexed man that, besides the municipal council, the peasants had agreed to this sentence. Next Sunday he wanted to go to the inn and ask the farmers himself, was his word.

In the meantime the innkeeper had successfully stirred things up against Pavel, and everyone, with the exception of Anton the blacksmith, was of the opinion that the lad was taking too much liberty and should be shown "it" once again. On Sunday Pavel came into the crowded inn, he did not take off his hat, in his company Arnost, whom he had befriended while on duty. He asked the only councillor present, an old man, and the round of those assembled, especially Peter, if he really had to pay, "yes, yes," was railed, "dog," it resounded to him from Peter. Then the unbelievable happened. Pavel calmly pulled an envelope out of his breast pocket, took a ten-guilder note from it and told the landlord to balance it and hand it over. No one expected this, malignity and disappointment spread, and then Pavel shouted, "And now I tell the town council and the peasants that they are all together a rag baggage" (Ebner-Eschenbach 1985, p. 148). A single outcry was the response to this outrageous insult thrown at the rich, the rulers by the least of the village. After the furniture of the inn had been smashed, heads beaten bloody, men flung out through the front door, Pavel, Anton and Arnost against the rest, Peter having wounded Pavel with a knife and the latter having lifted him into the air but then refrained from smashing him on the ground, the parishioner had a different position in the village than ever before, and even his worst enemies clenched their fists only in their trouser pockets.

Symbolic is an event that happened on the way home from the pub brawl. Pavel took home a dog bitten half to death by several mutts and nursed it back to health. "One against a lot too" thought Pavel. He remained a hard-to-reach but brave animal, and became a strong help to him in the last fight he had to fight with thuggish, blasphemous woodsmen. The winter passed, Pavel made another pilgrimage to the monastery to visit his sister, who made a sickly impression, the matron again successfully played the part of the nun keeping Pavel at a distance and swallowing the novice, and he made himself disappointed and at odds with himself, that he should have been more definite and persistent, on his way home, on which he – met his old teacher again, who had come to see him, and who was now no longer a teacher, and at last gave him his watch and six books, with the order not to forget how to read. The teacher disappeared with the train, Pavel went to the baroness to ask her to intercede for his sister, that she might be more spared, or else she must die, which the lady of the castle agreed she could not do, but gave him a field adjoining his sand-pit, which was officially certified. "The unheard-of fortune that fell to him from heaven, however, did nothing to diminish his unpopularity. No one begrudged him it" (Ebner-Eschenbach 1985, p. 179), the thieving did not stop, women and children cut off the still green wheat with sickles in the night and Pavel became crossly unhappy about it. The gift stirred up envy, the steward drew the baroness's unjustifiable magnanimity, and even the "spiritual lord" joined in this jeremiad, whereupon the old lady silenced him by saying that it was a fitting and not too magnanimous a gift for a good fellow hitherto neglected by fate. Her apparent change of heart had probably to do with the fact that Pavel was the brother of the girl whom she had chosen as her favourite, and in whom she already wished to see the future matron of the convent.

Winter set in with unusual cold, the most frequented path into the woods wound past Pavel's house, and the woodsmen came along on it. The cheekiest and most begrudging, Hanusch, provoked again one day by pointing to a roof truss already finished in the little garden. "It's finished, now you can start building the stable … Build it! Build it! Go ahead, the one you want to hire is already on her way … the one from the penitentiary" (Ebner-Eschenbach 1985, p. 191). The mother in the stable, that was too much, the sharp carpenter's axe flew through the air, hurt Hanusch's ear and Pavel jumped over the wooden fence, into the middle of the woodsmen. Hanush was the quickest to retreat, but already the dog confronted him and jumped at his throat. Pavel's "Back, Lamur!" saved his life. Grumbling, the lumberjacks went on their way. But Pavel began to imagine that he could forget both his early love for Vinska and all the suffering, and become a free man forever – "lonely and free".

How was his sister, Pavel asked, presenting himself once more at the convent, this time together with the baroness. "She has recovered," the matron told him. "Entered into the eternal light" (Ebner-Eschenbach 1985, p. 199). His pain mocked all consolation. Back in the village, he saw the parish shepherd sitting in front of Vinska's house, Peter had died in the meantime, and Vinska opened to him that his mother had arrived but had not allowed herself to be invited and was now waiting for him in front of Pavel's house. It is possible that at this meeting between Vinska and Pavel the latter suspected that she was filled with heavy but useless remorse for all that she had done to him, and which could never be repaired. A crowd of curious people accompanied him to his house to see the reunion between mother and son. She would not comply with the invitation to enter the house which followed his greeting. "I don't want to disgrace you, Pavel" (Ebner-Eschenbach 1985, p. 204). But he insisted and led her inside, telling her of her sister's death and that he had built the house for her. She, however, did not want to, said that he could not use her here, because all these years he had never looked around for her, and that is why she had only come to see how he was doing, but would return again. Not to the penitentiary, no, but to the hospital, where she works as a nurse. It must have occurred to Pavel then that his mother had once written him a letter, which he never answered, but tore it up, because Vinska had told him that he must not accept any letters from the female prisoner, because otherwise the whole family would get a bad name. The meeting of the two people was tinged with a quiet discord, for Pavel asked his mother, who had always kept silent, even in the trial before the judge, whether she had had no part in her father's crime – a question that pained her. In the end, however, she stayed with him, both hoping that even the most malicious would one day change.

At first glance, it might seem as if Pavel Holub's path from complete exclusion to a position of grudging recognition in the same village could be attributed to him and his individual efforts alone, relegating teacher Habrecht or other acting persons to a lesser position (cf. e.g. Friedländer 1985). From the point of view of correlative delivery and involvement, however, the situation is different. The inescapable interrelation between the social co-world and the individual is reflected on Pavel's side in the way that he completely internalises the role of outsider, thief and manslaughter abductee attributed to him, i.e. internalises the negative attributions and makes them his own in such a way that he wants to surpass them even further and demonstratively exaggerate himself to the outside world as what is attributed to him. This has a fatal reinforcing effect on the part of those around him, turning initial prejudices that could perhaps still be corrected by different behaviour (as a child he did not belong to the village community at all, but was an immigrant, unknown to the people) into rock-hard convictions. Habitualized action is rehearsed action – all human action is subject to the law of habituation – which proceeds in a typical way and is repeated and reinforced in the reciprocal reaction of people. If this process occurs again and again with some expectability, one can speak of institutionalization. This phenomenon is also understood from normativity and validity, as mentioned earlier. A community always feels challenged, threatened, and infiltrated by outsiders, and therefore devotes all its interest, suspicion, and finally persecutes them with hatred. Prejudiced, envious and ill-intentioned behaviour, as in the case of Pavel's fellow worldlings, has become an institution, has become entrenched, social and psychological processes intertwine uninhibitedly, the entanglement could only be changed by strongly intervening learning processes – as will then be shown by the tavern brawl and its accompanying circumstances. An old sociological insight coincides completely with this view: once a person has committed himself (in the context of opinions, attitudes and beliefs) to a position, this commitment itself becomes an obstacle against change, even if there are immediate counter-influences. With consistent support from parents, reference groups, or one's social class, change becomes unlikely (Berelson and Steiner 1972, p. 367).

The person becomes caught in the constraining corset of his or her own beliefs. These contexts, however, were embedded in objective conditions of a massively disadvantaged life situation, under which any action other than the one observed seems almost impossible. The fact that Pavel was able to get paid work now and then as a youth may have been related to the situation that incipient capitalism needed workers, and that entrepreneurs in those early days probably still had little regard for qualifications, age, etc. It is easy to see that with the common abstractions of transition to capitalism, declassification, etc., less comes to mind than in the literary account.

In the introduction to this chapter, the role of criticism was emphasized. The fact that a long-prevailing image, which judged the poet as preserving the existing order, sparing the role of the nobility, and, like Peter Rosegger and Adalbert Stifter, counted her among the "Habsburg Heimatliteratur" (Magris 1966, p. 153 f., note in Zeyringer and Gollner 2012), has been corrected to a considerable extent must be emphatically emphasized here (cf. Strigl et al. 2015; Strigl 2016). But how does the critique in the text here become sociologically significant? The subtle irony with which the matron is treated ("infinitely pious, infinitely apathetic"), the baroness who knits little jackets for the poor children in the village (!), the abysmally ambivalent Vinska, the priest who is clearly dispatched by the baroness, all of them and others testify to a critical attitude on the part of the author that can easily be seen as social criticism, even if it is not out for revolution. From the point of view of involvement and disengagement, we are dealing here with a perspective that questions everyday self-evidence (the world as it is "unquestioningly given" to us, as Alfred Schütz said), so that a way becomes visible how it is possible to take up a reflexive position outside of it, which can become the starting point for a different interpretation in turn. To this end, it is important to note that in a practical perspective, in the complete involvement in the everyday world, reflexive critique does not seem possible; it requires a theoretical perspective that has already left the self-evident. When a sociological analysis starts from literature, it must be clear that a text taken as a basis represents a reflection on an everyday world that has been deprived of self-evidence by the representation, and that the sociological interpretation only builds on this previous one, being free to choose the contexts of interpretation according to its own standards. Also, as already mentioned, Pavel's entanglement in his social co-world should be considered more closely. What makes the institutionalization of spiteful prejudice seem breakable are individual people who contribute to Pavel's ability to find a path, albeit after some detours and incursions, by which he can achieve what he has in mind: through possession and prestige, a correction of circumstances and of himself, a "new life." In the first place, there is probably the kind, humanistically minded, but somewhat inactive teacher

Habrecht. In a community, however, where he must keep secret, it seems especially from the priest, that he reads Lucretius and from this "de rerum natura," who's surprised? But the teacher does not remain the only person weighed against him. The blacksmith more often speaks out against the prejudices of others and becomes his friend, the baroness changes her attitude, the military comrade Arnost comes to his side, and after the death of his foster mother, even the foster father Virgil comes to him in a friendly way and offering help (to guard his new field against brigandage in the night). Last but not least, the sharpness and persistent vehemence of the slander and attacks seem to have receded, for the greatest adversary, the mayor's son Peter, has died; with his death Vinska also seems to have withdrawn her poisonous sting, and, what should not be overlooked, he has become one of the sturdiest men in a peasant community with the usual fisticuffs. Arguably, his successful integration, his participation in social life in a changed and positive way, is due to his drive and determination, but this would have been difficult to achieve without a change in external circumstances (ownership of land) and without the support of other people.

6.2   "The Hagestolz" (Adalbert Stifter)

To some, this narrative is considered Adalbert Stifter's most beautiful, and the character of Viktor appears as an "ideal specimen" (Gollner 2012, p. 327) of beauty and morality as drawn by the author. To me, however, the figure of the old Oheim wants to seem just as memorable, but more psychologically complex. Now what could a hag's pride be? An old lexical information comes close to the character Adalbert Stifter drew:

> Diecmann in Spec. Gloss. Lat. Theot. derives it from hag, house, and stallt, the middle word of stellen, and explains it by persons who make themselves comfortable for a voluntary solitude, who, as it were, shut themselves up in their house. This derivation is made probable by the fact that haistaldi or haistoldi occurs in du Fresne for people who live in their house; a proud hag is also called Einstöding in a similar way in Swedish, and Einstädingur in Iceland, from ein, alone, and stä, to stand, as it were a lonely person. ("Oekonomische Encyclopaedie" by Johann Georg Krünitz 1773–1858)[3]

The narrative begins with a bright sunny day in spring, a group of young men wandering towards a mountain height "and down in the plain looked the towers and

---

[3] Source: http://www.kruenitz1.uni-trier.de/

house loads of a great city" (Stifter 1951, p. 582). They are obviously young people, gripped by romantic liberalism; for the state they propose "the greatest freedom, the greatest justice and the greatest forbearance" (Stifter 1951, p. 582); the enemy of the country, whoever that may be, must be crushed; and while they speak of the great, they think that only the small is happening around them, not yet knowing that the great is also hidden in the small. They are high-spirited, one of them conspires that he will never marry, one hears the cock crowing from afar, hears bells ringing from the church tower, there are still no townspeople here in their summer homes, a rural idyll prevails. At the end of the walk, when they part in the evening, Viktor turns out to be the one who never wants to marry and is completely unhappy. In good contrast, Adalbert Stifter has an old man sitting in the sunshine on the bench in front of his house on the same day of the youthful hike, far "behind the shining blue mountains," on an island in the middle of a lake. "The old man sat by the house, trembling with dying" (Stifter 1951, p. 585). He does not want to be seen by people, never had a wife and children, in the house it is silent, it could be that he has already thought of suicide, but he is afraid of it, he is deeply lonely. Adalbert Stifter has carefully named the first chapter of the story "Gegenbild".

Viktor, however, is to move away from the small, clean house with the shiny windows in the wooded hills, where he has lived with his foster mother Ludmilla and his foster sister Hanna since childhood, and take up a civil servant's post in the distant city. Away, then, from this cosy idyll, where everyone loves each other and treats each other with courtesy and respect, away from this sheltered world, for whom "the city" already means hostile terrain, where man is endangered, above all morally. But before he embarks on this journey, he must visit his uncle, who demands this visit by way of guardianship, and on foot. Victor's parents died when he was young, he has no memory of them, the father was a good and kind man, but economically unsuccessful, leaving only a small property burdened with debts, to which the guardian seems to lay claim. The farewell ceremonies last for almost 2 days, the foster mother and Hanna have put together for him an abundance of underwear ("linen") as well as the day's garments, the suitcases have already been prepared, which he now packs, watched carefully by the dog, an old spitz; the foster mother has put money aside for him (bypassing the guardian), which he refuses to accept out of excessive modesty, Hanna has made him a small purse and a splendid wallet as a farewell gift; Viktor laments to his foster mother that he is alone in the world and will never marry; she comforts him for the future and wants to direct his attention to the guardian's daughter, Rosina, who, however, is only 12 years old; the guardian comes with family to the last supper, pronounces admonitions and rules of conduct, brings a letter to the uncle as well as letters of recommendation for the town, and at last Viktor and Hanna, the foster siblings who have grown up together for years, also confess their love for each other. The next morning, Viktor leaves home and family and sets out to visit his uncle, the brother of his long-dead father.

It is hard to imagine a more perfectly integrated social world, in which each has its place and its task, in which love, contentment, modesty, closeness to nature and a faithful understanding of the world form the foundation of the whole. "The happiness of the place and of the persons is joined of diligence, cleanliness, order, and righteousness. In such a framework all love all" (Gollner 2012, p. 327). It is a model of perfect integration, which does not seem to be endangered in any way. That a life path out of such circumstances, in which anyone who is weak or has less advantageous starting chances than others is faithfully taken care of, could fail anyone, belongs to what is quite unlikely in the world. That is why Viktor's demonstrative dissatisfaction with life, his feeling of loneliness and his refusal to marry seem to be imposed and do not quite fit into this picture of over-harmonious life, if it is not to be taken as a late-pubescent romantic surge of emotion. This young man, distinguished by beauty, innocence and goodness of heart, is now to meet his uncle, his distorted image, as it were.

It was a long hike, on the third day old Spitz caught up with him, "terribly emaciated," on the eighth day he came to an inhospitable region, met a man in the gazebo of a village inn, to whom he said in conversation, "I really want to go to the Hul." "To the Hul? – You'll get a bad reception there" (Stifter 1951, p. 608). Inquiring his way further, he came there and had an old man row him across the lake to the "Klause" on the island, which had formerly been a monastery; this old man also meant to him that he would be badly received there. And it was so, the reception at the closed garden gate was unfriendly and cold, on handing over the letter the old man said: "Your guardian is a fool and a limited man" (Stifter 1951, p. 615); whether he had really come on foot he doubted at first (this with Viktor, "who had heard no inconsiderate words in his life"), finally he wanted Viktor to drown the dog with a stone on a string in the lake before he would let him in. As Viktor refuses, the garden door in front of which he stands is not opened, and he makes camp for the night with the dog in a bush, when Christoph, the servant, calls him to dinner: his uncle has been waiting a quarter of an hour. "The master will certainly have already begun to eat; for he has his appointed hours and does not depart from them" (Stifter 1951, p. 619), for reasons of health, as he says. Viktor goes with the servant to the house on condition that nothing happens to the dog, the rich meal is served by an "old woman", but the benevolent heart he had wanted to bring here, says the poet, was choked to him. After dinner the gruff and unkind landlord took him to his room and locked the gate in the hallway "and there was the silence of the dead in the house" (Stifter 1951, p. 622); but Viktor now remembered that he had seen only three people in the house that day "and that these had been all old ones." Despite Viktor's friendliness, his uncle spreads an annoying mood, speaks only in a commanding manner, constantly blocks the doors, even during the day, looks immensely haggard and dilapidated, and the completely grey hair had "never, since it grew, been stroked by a loving hand" (Stifter 1951, p. 626). After 6 days of the stay Viktor could stand it no longer, asked the uncle why he had had to come without receiving an answer, and told him that he wanted to leave again, which the old man forbade him to do. "So I am a prisoner?" "If you call it so, and my institutions make it so, then you are one" (Stifter 1951, p. 634). No other character has been drawn so harshly by Adalbert Stifter, so that the later change must come down on Viktor like a landslide. A subsequent suicide threat by Viktor that he would throw himself off the cliffs into the lake was laughed at, but at last he agreed to stay on. Doubts came to him as to whether the old man was really so hard, or else an unhappy old man. Rosalie, the housekeeper and cook, was old; the grandfather's three dogs were old; the fruit trees were old; the stone dwarfs in the garden were old. "So the two people lived side by side (…) who should have been closer to each other than any other people (…), Viktor the free cheerful beginning, with gentle flashes of the eye, an open place for future deeds and joys – the other the degenerate, with the intimidated look and with a bitter past in every train" (Stifter 1951, p. 638). The uncle trusted no one, he shaved himself so that no one would cut off his neck, he locked up the dogs so that they would not eat him at night. In Alfred Adler's sense, the Hagestolz is a failure because he lacks a sense of community and a stake in it (Adler 1979, p. 16). But slowly they began to draw closer to each other, may have been pity in the case of the young and a hesitant departure from intentional harshness in the case of the old, they talked to each other, especially during meals.

The actual turning point, however, came when Viktor, the time remaining to him until taking up the office had almost expired, was confronted by his uncle with the request (!) to voluntarily (!) stay a few more days to see "whether we lived well together", and at the same time showed him a letter from the office with the confirmation: "Leave for an indefinite period". So he was able to do that, despite his seclusion, and at the same time the problem-free extension of the vacancy became the outlet for a gloat: How important must the young man be for the office if he was given leave before he had even taken up the post? He should learn, get to know the world first and – marry, not immediately, but while he is still young. It is an outcry: "O Viktor, do you know life? do you know that thing called age?" (Stifter 1951, p. 647). And then follows a poetic résonnement on old age:

Life is immeasurably long while one is still young. One always thinks that one still has quite a lot ahead of one and that one has only gone a short way. That's why you put things off, put this and that aside to do later. But when you want to do it, it is too late, and you realize that you are old. That is why life is an incalculable field when you look at it from the front, and it is barely two spans long when you look back at the end. (…) Everything disintegrates in an instant if one has not created an existence that continues above the coffin. Those around whom sons, grandchildren and great-grandchildren sit at their old age often live to be a thousand years old. (Stifter 1951, p. 647)

With his death, everything that he had been would fall away, therefore Viktor would have to marry, and the continuation of life in the descendants would become the focus of the whole outpouring. Here one could speak of a "transference", as is common in depth psychology, especially in psychoanalysis, which refers to the process by which a person unconsciously transfers and reactivates old – often re-pressed – feelings, affects, expectations (especially role expectations), desires and fears to new social relationships. The uncle seems to transfer the desires for marriage, children and family to Viktor.[4] The old man is not a comedic figure, he is serious and he suffers from his childlessness, which made him cast his eye early on Viktor, who should have everything one day. But with him, too, what was said above in the case of Pavel's opponents is true: The forced corset of one's own convictions misplaces almost every way out of a misery – also perceived in this way – for the person concerned.

Then the old man told Victor that he had a debt-free estate, together with money and securities, which he wanted to bequeath to him after he had travelled the world for 2 or 3 years; he would rather not see him in an office: "I mean, you should be a farmer, as the ancient Romans were,[5] (…) If you are wise, it is good; if you are a fool, you may regret your life in old age, as I regretted mine" (Stifter 1951, p. 648). After a convoluted backstory that finally reveals how the uncles relationships failed and failed and why Viktor's father fell, he finally wants to give the boy the papers, but the boy – refuses. The last thing Viktor sees in the old man's eyes before he pushes him out the garden door is – tears. Back home, he tells his foster mother that his uncle is "a splendid, excellent man". Viktor returns from his journey to "foreign lands" after 4 years, the uncles papers are now with the guardian, and not long afterwards he stands with Hanna "for eternal union at the altar".

---

[4] That fears and unfulfilled desires stem only from childhood, as Sigmund Freud assumed, is no longer universally held today.

[5] The theme of the Roman peasant who cultivates the land and gathers his strength is significant in Adalbert Stifter; it also reappears in the story "Brigitta" (Stifter 1951, p. 559).

The uncle did not come to the wedding. "He sat all alone on his island; for, as he himself had once said, everything was too late, and what had been neglected could not be made up for" (Stifter 1951, p. 660). However, the fact that despite all the harshness and grief, the old man also has a conciliatory side, is echoed at the very end. The biblical parable of the fig tree (Luke 13:7) is defused: "The kind, mild and great gardener does not throw it into the fire, but looks at the fruitless foliage every spring and lets it green up every spring, until once the leaves are fewer and finally only the scrawny branches stick up" (Stifter 1951, p. 660).

The uncle's disengagement from social relations during his life course is highly determined by self-activity; the environment's disengagement signals are simple reactions to his behavior. However, this self-chosen isolation, in which almost no social and certainly no political and cultural participation takes place, can apparently only be successful in the long run because the uncle has already created the economic foundations he needs for this beforehand – in this respect the social situation is relevant. He has bought the whole island including the hermitage, he can afford to employ servants, and he can bequeath an apparently handsome inheritance to Victor. The cold and stifling atmosphere in his uncles hermitage, the frosty mood he spreads, could be clearly described in scientific terms, sociological and psychological repertoire could be used, and yet the literary representation in the narrative adds something qualitatively different that could not be found in any abstraction. There is also another reality than the theoretically measurable one (Hans Blumenberg). It is similar with the description of the domestic circumstances surrounding Ludmilla and Hanna, which can of course be grasped as a romanticised family model and could be associated with the author's "addiction to harmony" (Helmut Gollner),[6] with his dream of a happy life that was never his. But the fact that these very descriptions appeal to the reader's emotions, even to the point of embarrassment at some of Viktor's behaviour, is due to the nature of the narrative and its own access to a fictitious reality, and does not disappear by the explanation that views and mores have changed and that today we look at the world more soberly (if this is true). Nor do I want to let stand the fact that Adalbert Stifter uncritically conjured up a harmonious world instead of criticizing it, since the painting of a happy and undamaged world can still express itself as unease at a damaged one. It should also be present to a sociologically minded person that not only do many images and the expressions that signify them move within conventional forms, but that conventional forms of expression themselves in part generate the

---

[6] According to the testimony of Karl Kraus, Adalbert Stifter was called by the old Austrian writer, musician and pedagogue Josef V. Widmann the "peace of mind founder" (cf. Kraus "Die Fackel" 418, 57).

content to be represented. This leads to the familiar consideration that artistic output, e.g. literature, probably never appears to later times in its original and then perhaps unambiguous form, but is always laden with an ingredient, perhaps even enriched, which subsequent interpretation added. Will anyone ever be able to read Miguel de Cervantes' "Don Quixote" again in the same way as the first time, when, after reading it for the first time, he has then read "Pierre Menard, Author of the Quixote" by Jorge L. Borges or seen the four-part film "The Story of Don Quixote of La Mancha" directed by Carlo Rim and starring Josef Meinrad?

For both pieces of literature discussed here, I would like to assert the following thought: due to their traditional elements originating from the nineteenth century and earlier, they appear to be a result of the past, but thanks to their originality and the features updated at that time, they can also be read as an aid for enriching an already existing picture of the past, which science so circumstantially does not have at hand. In the assumptions of understanding sociology, the everyday world is given to people as "self-evident". Reflected self-evidence – in art the fictitious representation, in science the theory – is no longer self-evident. To even want to name the self-evident, or even to describe it, means nothing other than to deprive it of its character (cf. Blumenberg 2010). When sociology makes literature its object, it reflects the result of a process of annihilation of the self-evident in the everyday world that has already taken place, and thus continues the endless process of interpreting the world according to its own standards. In doing so, it may sometimes succeed, despite its self-commitment to abstraction, in mirroring the living of the self-evident.

About Telling, About Being There and About Structures

7

7.1   Storytelling in Everyday Life

The following considerations are not about literary narrative, but about narrative in everyday life, which is conceded to form "an essential element in our understanding of reality" (Butor 1984, p. 53), and which Hayden White has said is a "pan-global fact of culture" (White 1981, p. 1). Michael Scheffel, in a review of narratological research, gives the following structural features and pragmatic functions of narrative[1]: The transfer of events into stories, whereby temporal facts can be organized and brought into a meaningful context, in that through remembering, visualizing and imagining sequences of events, the explanation and thus cognitive coping with spatio-temporal data occurs. Every narrative represents a communicative act that establishes, multiplies and differentiates social relations by way of different narrative forms and larger or smaller stories. More generally, narrative enables people to found and maintain communities, but also to differentiate the identity of individuals and collectives. These moments fit neatly into the process of involvement and disengagement; they are (among others) constituents of the social world. In order to illustrate these abstract considerations somewhat more vividly, I will describe two model cases of narrative locations, which I know on the one hand from my own experience, and on the other hand precisely from narratives, and give brief examples of narratives.[2]

---

[1] https://www.uni-bamberg.de/fileadmin/uni/fakultaeten/split_lehrstuehle/didaktik_deutsch/Daten/Material_Brendel-Perpina/Erzaehlen/MichaelScheffel-TheorieUndPraxisDesErzaehlens.PDF

[2] Source: subsequently recorded interview transcripts.

© The Author(s), under exclusive license to Springer Fachmedien Wiesbaden GmbH, part of Springer Nature 2023
A. Amann, *Living - Participating - Growing Old*,
https://doi.org/10.1007/978-3-658-39681-7_7

Narrative weaves through our everyday life in great density, it creates small communities, inspires them, keeps them alive, gives those involved the feeling and experience of being there, of participating in what is seen as important – without much thought about why it might be important and for whom, and for what reasons. I dare say there are "village" narratives, even in cities or their small-scale substructures, made up of hundreds of stories that have been and are being told. Each narrative serves to reassure, correct and add to what is already known. All those involved find themselves confirmed in their stories, learn something new, are corrected, and are inspired to tell others. Narratives are the glue of small communities, it binds twofold: each narrative in the context of what has already been told and all those involved in the community. Those who narrate bind themselves in and are bound in, those who never participate disentangle themselves, are disentangled. Not from the community at all, but definitely from the round of tellers. There are the experts of stories, for all sorts of reasons, they enjoy a certain recognition. There is probably also a "consciousness of ability" (Scheler 1954, p. 248) that plays a role here. Not infrequently the older ones belong to them, who remember stories from past times, which are accessible to the younger ones only as told, but to the older ones from their own experience. This, too, creates a common ground, one of a very specific kind, which can be based on the authenticity of the reported experience. When I report on this narrative, it is not a romantic mourning over the suppression of old-fashioned customs and primitive life as a moral instance that is playing itself out here; rather, community is understood as a process that is inconceivable without society – the two alleged opposites have always been nourished by the erroneous idea that one really existed without the other, or even that society developed out of community.

7.1.1 The Village Regulars' Table

"The men are on their way to themselves when they are together in the evening, drink and talk and mean," says Ingeborg Bachmann in the first sentence of her story "Unter Mördern und Irren" (Among Murderers and Lunatics) (Bachmann 1996, p. 159). However, the only reference that can be made from the Stammtisch to this story is the one that refers to the exclusivity of the round, but not to the lunatics and murderers in the story. The village determines the outer framework, the living conditions are not only clear, they are also more or less exactly known to all. The men in this case sit in the inn 4 days a week, between 9.30 and 11.00, the only women to be seen at all belong to the service staff, at most once a wife comes to fetch her husband, who is on crutches. In this case, however, she does not sit down, but waits until he has paid and they can leave. Remnants of patriarchal structures and world views. Yet among the visitors there are some who always come, unless serious reasons prevent it, and others who show up sporadically. The very fact that the regulars' table takes place on weekday mornings gives us an inkling that the majority of its members are pensioners. The most important activity, to which everyone is more or less permanently committed, is telling stories. They talk about things that have happened, current or past, about events, relationships and circumstances. By telling stories, they transfer what has happened into stories in which the events not only follow one another, but also diverge, in that one story or a part of it becomes the starting point for another. The rough demarcation of themes can best be determined by everyday areas. These men's rounds are dominated by working life, community politics, life stories and circumstances of individuals (often of the recently deceased), and technology and crafts. In the topic area of work, the road renovation is discussed, one knows the foreman of the company responsible and reports on his tale of the difficult geological conditions, another, formerly an excavator driver, expertly describes a newly used machine that mills off the asphalt and immediately prepares the material for further use, still another knows to report on the calculated costs in the municipal budget, which in turn becomes the occasion for a next tale about the municipal finances. The narrative path leads almost seamlessly from finances to community politics and the problem of second homes, about which a heated argument arises, in which some village members are listed who have sold building plots on which these houses now stand. One again knows very well how and why a farmer's son who has sold such a plot of land is about to put the money through uselessly. Two camps emerge as to the correct use of a dialect expression, a generally unanimously shared opinion is not reached, but a new tale emerges about the allegedly comprehensive knowledge of the old valley chronicler, whom almost all of those present still knew, and to whom a commemorative plaque was dedicated by the community. The first two ask for payment and leave, a remark is made about one of them, but it is not answered or responded to, the conversation dies down and the group disperses. It is fair to say that the exchange of knowledge and opinions, the agreement and also disagreement that are expressed, the demonstration of knowledge of place and fact in the telling, fortify the images that people have of the world and of each other as well as of themselves. Yet some not entirely unexpected insights are also of concern. There is no shortage of prejudice about asylum seekers, moonlighters and intellectuals, nor of ostentatious dissemination of misinformation; there is no shortage of sarcasm and irony. Half-truths and exaggerations occur (a shadow on a lung X-ray becomes a carcinoma in its final stage after three or four further narratives), and for some of the narratives can apply: The longer life lasts, the more persons and experiences from earlier times become anecdotes. If they are told over and over again, they sinter out into monuments.

7.1.2 The Parish Women's Group

Here the setting is not the village, but a city, with its small social co-worlds known to the individual women, but also its many anonymous areas. The social space that creates mutual acquaintance and knowledge of the surroundings is the parish. The women belong to the parish, they feel they belong and they are responsible for some things – for example, for home visits to elderly people for whom they do the shopping, organize a coffee and cake afternoon once a month, and more, all on a voluntary basis. The meeting of the women's group takes place every 2 weeks on a Monday afternoon, between 16.00 and 18.00, it has a quiet obligation character. Those who stay away should let the others know in advance. The group consists of six women, the number is more or less always the same. The main themes of their stories are their auxiliary activities and the people involved in them, parish and church matters, family and children, and household matters. Four of the women are between 40 and 50, two are retired, and the younger ones are all employed part-time. Last week, among others, an old lady was visited who has been a widow for 15 years, and who, now that her daughter, whom she had tried with all her force and refinement to keep, has also married and moved away, lives all alone and almost without address. It had not been good for the daughter, she said; it was almost a miracle that her now husband had waited so long, because the marriage had been put off again and again. That was true, but the old woman had fallen into a deep hole, she could literally be seen withdrawing from life. In the case of the woman in Lion Street 12 it was quite different, she still had her son living with her, who was now almost 50 years old and who was taking very good care of her. Oh yes, that's the one who has already had three of those he wanted to marry run away again because they couldn't stand his mother. This week we are to visit an old man whom no one likes to go to because he is quarrelsome and hot-tempered. It would be good if the same helper always went to see him, that would build up trust, but everyone wants to keep the visiting contacts to a minimum. Even a conversation between the priest and the old man did not help. One has to take note that some people become difficult and cranky when they get old. Besides, the priest, who has only been here a year, is more sociable and less controlling than the old one, who could hardly be made to like anything, and who kept up some of the old traditions, even though hardly any of the parishioners participated in them. But one had to be careful, the new pastor made an impression on the young women, M. Z. was to be found conspicuously often in church, at least in comparison to before. One of the participants was asked about her sister, what she had actually had, they had heard that she had had to go to hospital; after some hesitation they were told that she had had an abortion, which led to another story about a couple who had wanted a child for a long time, but so far had remained childless. The schedule of volunteer services is then discussed, it would be urgent to expand the circle, but it is difficult to find interested people. Finally, they decided who would bring the cake to the next meeting. Storytelling among women is different from that among men. I dare to assume that in these narratives among women, the perlocutionary speech acts are less frequent and the personal experiences, including those that affect them, come more to the fore.

From the village table and from the parish women's group we can generalize: In all "villages"[3] stories are told, stories that sound somewhat different from the mouths of each individual person – the accents change, the roles of the tellers change, one person omits what another emphasizes – but which all together serve to constantly weave, supplement, correct the fabric of the whole community. This story-telling ensures the embedding of the events of the day as well as the salient markers of incidents long past. However, we have to take note of the fact that it is almost never a matter of enlightening discourse, and that the securing of one's own place happens to a large extent through the passing on of prejudices and stereotypes. The fabric of events in a community is not closed, and in principle does not lock people out; but since it is the place of self-assurance and self-understanding for those involved, it sets up barriers and boundaries for those who come from outside, layer by layer, the more they try to penetrate inside (which is why you will very rarely find strangers at regulars' tables in the village – where these still exist). This is much more than the "lack of information," as communication theorists claim; it is the density and tenacity of habitual thought and feeling, the softening of which is experienced as alienation. "Who is at home here, the tourists or me?", a farmer's wife is said to have asked the mayor when she was accused of damaging the cross-country ski trail with her dung cart.

---

[3] I have met people in all the big cities, from Vienna to New York and Rome to London and Paris, who cultivated their small communities, who knew people and their stories very well in their small circle (in the midst of all anonymity), and thus hardly differed from those who are said to live in the countryside or in the provinces.

7.2   A Little Philosophy of Old Age, Everyday Life and Interrelationships

In public, especially political discourse, ageing and old age have become a shadowy topic – except when it comes to pensions and care. In most European countries, the wave of nationalist self-delusion is sweeping away all important issues on which the future hangs, including old age. Politics and academia have slain utopia. Above all, the "positivist" social sciences have killed it off. Today it is considered not only unscientific but downright insane to talk of utopia. This decline is inconsequential for assessing the role of utopias when they were effective – in the seventeenth, eighteenth and nineteenth centuries. It is not inconsequential to the question of what the consequences will be. The complete disappearance of this kind of thinking, which had triggered progress in the past, progress towards a better society, may after all mean that social progress of this kind will also decline – as an ideal and as a practice. Always, utopian thinking has been able to do three things: it gave strength to accept the unchangeable; it gave courage to change what should be changed; it gave wisdom to distinguish between the two. In this way, this thinking was a scout for the better, a bridge between the present and the future, a driving force and trigger of social progress. Imagination, which is called for today, would have to name what is currently bad, and it would have to do so very precisely. It would have to find words that correspond to the matter at hand and stand up to the flood of empty words. "Harmonization of pension systems", "equalization of benefits", "cushioning of hardships", or similar phrases do not name, they disguise: that taken away from some and not decisively given to others, that what someone is supposed to get will be less than before. Besides, they come from accountant and mechanic language and are not appropriate to life. Fantasy that ventures into utopia is not in the business of picturing in detail a better society of the future. Nor does it have the task of showing practically the way to get there step by step. It would have to look critically at the present conditions, to justify the criticism, and to state clearly the intention that is being pursued. It is the intention that counts. That is why it was once spoken of "wish-space" (an expression of the historian Alfred Doren). It lies where a single wishing is justifiable. One that is directed towards a social constitution that does not oppress and stultify people, does not exploit them and turn them into the kneading mass of ruling interests. In other words, a form of organization in which individual life is in the foreground. This requires constant questioning. It must be radical, even revolutionary. Questions must have the effect of a sting and pierce complacency and the always familiar. Utopian thinking does its service when everyone gets the feeling that crusts and ramparts are being broken through. It is a satirist and sharp critic who cannot be muzzled. It denounces mistakes and relentlessly makes all aware of their responsibilities. Utopian thinking is the mother and father of the better (Amann 2004).

Dennis Gabor, the famous physicist, following the French philosopher, Julien Benda, at the time of the Cold War, thought about the silence of those who could have known better. He called it the "betrayal of intellectuals". This manifested itself above all in their inability to make plausible and stimulating proposals for solving the mounting crises in almost all areas of life.

> We have to "invent the future", he said. Many would probably still refuse to accept this thought today. Nevertheless, it is correct. Very different thinkers, such as the Dutch forecaster Fred Polak, the philosopher Ernst Bloch, or the American publicist Alvin Toffler, with his idea of "anticipatory democracy", have shown what is at stake. Always, the designs of future images have provided inspiration for Western civilization to a great extent. If this necessary capacity for anticipatory design dries up, the road to decline is paved. New futures must first be created in the minds and hearts of people. Then the power of these ideas will help to attract and change the world. (Amann 2004)

We seem to have completely forgotten what Baruch Spinoza and Albert Einstein took for granted: to form our intuition, which is based on empathy with experience and represents the highest expression of the human spirit. This would be a contemplative knowledge in which sensuous imagination and conceptual reason are united. We have direct and indirect experience of the concentration of enormous financial and military power, of the indoctrination of advertising, of the surveillance of people and the gradual erosion of their civil and social rights that goes with it. We have experience of the intimidation of thought by the never-ending sensationalism of the media, of the constant influence of schools, radio, TV and the press. We have experience of the gagging of our imagination by the constant avalanche of phrases concerning our own affairs. Why is it that so few have a flash of insight, why does a sense of other realities return with such difficulty? Imaginative thinking that does not avoid the utopian, or even defames it, must begin with such questions (Amann 2004). These ideas can be combined well with considerations of the everyday world, integration and age.

All acts of the individual that lead to integration or disintegration are the result of socially conditioned individual decisions or self-selections. Even if the patterns of long-lasting social integration processes seem to have the rigid texture of old wood, they are always subject to the lively micro-changes in daily coexistence. Sociological research on everyday social integration now seems to me to have recently undergone a methodological drubbing that manifests itself in an increasingly rigorous formalization of the processing of qualitative material obtained in interactive survey procedures. The hermeneutic process is forced into a kind of corset, even if this has highly diverse architectures. One consequence of this is that the social conditionality of everyday decision-making practices, however individual they may appear, loses its meaning. Everyday life is not only the individual, but at the same time also the supra-individual. It seems to me that this is only one facet of the fantasized identity crisis of sociology that was already being talked about 20 years ago. In this matter, I profess to be a traditionalist and believe very much that there will be something like a survivable sociology, which will prove itself the better the more precisely it knows its limits and seeks its salvation neither in the quantifying tornado of data nor in the structure-forgetting interpretation of micro-moments of social events. If such a sociology is capable of imparting anything, it is a kind of perceptual competence for social processes. For despite all the digitalisation, people will not be saying goodbye to curiosity about social events any time soon.[4]

Anyone who writes about everyday life, as I do here, or who speaks about it from a reflective distance, must be careful not to smother its so-called banality in the arrogance of an erudite sophism. Two qualities cling to everyday life that are significant: Its monotony and compulsiveness, which make it irksome to us, but also the fact that only from it do we grow the strength to cope with it spiritually. "I despise the daily because it is always absurd," Johann W. v. Goethe has recorded, but against this it is also to be objected that everything which we call spiritual life, and which often comes along completely detached from everyday life, cannot come from anywhere else than from this everyday life, in which the trouble is imposed upon us to keep ourselves in existence, to remain connected with life, and to develop our forms of life constructively. What else should have driven people to search for the good, the noble and the beautiful than the experience of the opposite in everyday life? In Sect. 1.1.3, in connection with Helmuth Plessner's conception, the concept of the limit or border was particularly emphasized. Here his phenomenological conception of limit can be translated into an empirical-factual one. In order to exist as human beings, we need boundaries; without them we would be poured out into nothingness, we would be formless and groundless. This is true in a material and immaterial sense; our limits are of a physical and psycho-spiritual nature, they change in the course of our lives, and they can be described in a general way by the concept of scarcity (cf. Sect. 2.2). Our powers are limited, our influence on others is limited, our capacity for knowledge is limited. But space is also

---

[4] To the best of my recollection, Wendelin Schmidt-Dengler once said something similar in a lecture on literary studies.

limited, material resources and energy are limited. Philosophically, this could be captured in the notion that a person's realized possibilities are revealed by what limits them, by what becomes visible as the balance between vigorous outreach and harsh restriction. The notion of balance, however, seems to me to be more important than ever in the face of the ongoing destruction of the earth; it could provide a core approach for further research, far beyond the common thinking about balance, which sometimes seems hopelessly outdated.

Albert Camus said that the best way to get to know a city is to try to find out how people work in it, how they love in it and how they die in it (Camus 1950, p. 7). From the perspective of everyday life theory, this can be applied to any place where people live together, where they have set up their lives. And yet, there is something missing from this idea when we direct it to ourselves and to getting to know a place, a community: staying long enough to allow ourselves to change. The Latin saying "tempora mutantur, nos et mutamur in illis"[5] hits this nail on the head, because a change in circumstances is accompanied by changes in ourselves (and vice versa), in binding and unbinding this double-sense process is reflected. Eavesdropping on people's stories, as happens in everyday sociological surveys, writing down sections of the narrative fabric, is an act of approaching the experience of others. Now, approaching the experience of others is different from that of a thing; an old cupboard, a house, a church. The experiences of individuals are indivisible (which is why sharing also uses the form of stories) and referred back to themselves in the sense of remembering and interpreting in advance. The act of approaching the experience of others requires in-depth research, which produces closeness, and the synthesis of the many, which produces distance; it requires the right moment and the long duration. This was probably also reflected in Bronislaw Malinowski's recommendation to young ethnologists that they should pitch their tents with the people they wanted to study.

In another terminology, the idea refers to the social world already mentioned and the fact, essential for it, that participation in old age always takes shape through a multitude of conditions. Before I turn to this topic, the question should be clarified as to what is philosophical about this small philosophy of integration in old age.[6] By philosophy I want to understand not so much an initial wonder, the θαυμάζειν, as has been attributed to it as a characteristic since Plato, and which likes to emphasize the moment of wonder, but a thinking that insists on a radical

---

[5] Created in the sixteenth century after the verse "tempora labuntur tacitisque senescimus annis…" from the Fasti of Ovid.
[6] Since I cannot claim to be a professional philosopher, the only answer that remains is the kind that a dilettante would find appropriate.

always-on questioning. Thousands of statistical correlations exist between age-related factors that must be constantly re-extended and increased in complexity in order to approach meaningful insights. With philosophical impetus, this thinking has in common that it keeps systematically questioning both the results and the methods by which they were obtained. A simple frequency distribution about attitudes or behaviors (e.g., 41% of the adult population approve of the current government) are simplifications of contexts that border on shamelessness. It goes, as I emphasize in the preface, beyond simple opinion, and seeks to assure its tenability (which, again, is not so far from Plato, if his distinction between δόξα: mere opinion and ἀλήθεια: truth is taken into account). However, what has to be regarded as unambiguously characteristic in present understanding of scientific thought is the moment of empirically demonstrable tenability of conjectures or hypotheses until their refutation. All scientific knowledge is fallible. Under this assumption, the situation of human beings and the conditions of their integration in the social world as they grow older present themselves as multifactorial, for no single cause, for or against the success of integration, can be considered solely responsible. As is so often the case, only the most comprehensive examination possible will provide any reasonable insights.

7.2.1 Thinking in Contexts

According to the concept presented here, participation is a basic element of integration. It is also necessary to consider the perspective of participation rights and obligations, including the conditions for their realisation at the societal level through the provision and design of opportunities. Here, in the planning and design of many initiatives, it is crucially neglected that participation processes develop in a gradual character and need time. It starts with the right information to the right address, continues with participation and co-decision-making, and can finally culminate in self-administration. I have chosen this point of view with care, because it demonstrates a problem that touches directly on our subject: The difficult, and therefore perennially neglected, question of how our knowledge must be constituted in order for our practice to adequately shape conditions. This is the fundamental question of all politics, and science has the task of saying something about what we can know and what this knowledge should look like.

Individual research findings that only refer to selected and specific indications, such as the fact that 28% of female pensioners living alone are on the poverty line or in poverty, is a fact, but says nothing about its context. Unfortunately, all too often in the media discussion only such headlines are used. In fact, much more complex contexts should be made known. If we know that the greater proportion of these women have low formal education and employment careers that have been interrupted or broken off several times, we already know something more. If we could show that they have also worked in relatively low-income industries, that many of them have impaired health, that their previous social networks have shrunk after, for example, the death of their husbands and the marriage of their children, that their social resources may be weaker than those of other groups, we would already know much more. Especially with regard to the question of whether and what they could change about their situation themselves. If, in addition to a multitude of other such findings, there were also empirically verifiable insights into the constellations of interests and mechanisms that lead to women being systematically disadvantaged, if we were able to specifically name the groups and their interests that repeatedly bring about these conditions, then we would gradually be in a position to speak of adequate knowledge. Appropriate has to do with more comprehensive, ideally with holistic knowledge. That such knowledge means an ideal with which we can only ever work approximately is clear. But the crucial point is that we must always try to consider as many aspects as possible in any design task.

At the beginning, I used the term "practice" in addition to the term "knowledge". Now this practice is subject to the principle of a high degree of division of labour both in the bureaucratic organisations and in the political design work and thus to an often fragmented division in competences, powers and responsibilities. One might think here of the opinions of a wide variety of offices, bodies, party organisations and specialist groups on the very first recommendations for the Federal Plan for Senior Citizens, which was later implemented in Austria (2012); what is still striking about them today is how different interests, divergent values and, above all, how different, sometimes even contradictory, knowledge they were borne by, with party-political perspectives not infrequently dominating expertise. From an everyday practical point of view, it could now be said: That's just the way it is; in a sense, that's our fate in a democratic polity. But I think this answer is wrong, in a twofold way: Fragmented and contradictory knowledge is not justifiable in the face of a possible more comprehensive knowledge, and restriction to partial competences is formally justifiable, but in the sense of a desired common good – e.g. participation and quality of life for as many as possible – not very fruitful, sometimes even counterproductive. This is where my postulate of networked knowledge comes from. Using two examples, I will try to show what such networked insights might look like in the context of participatory reflection.

Education

All of the following correlations should be understood in terms of proven correlations. There is a positive effect of continuous mental stimulation on the maintenance of good health. Mental training has a positive influence on intellectual abilities, for example by reducing or reversing memory loss. In any case, learning leads to a change in brain structure. Higher education reduces the risk of dementia and mortality. Furthermore, participation in continuing education leads to social integration and reinforces a positive social image of old age, increases physical and mental well-being, improves the anticipation and processing of critical life events, and has a positive effect on civic engagement and volunteer work. Education in old age contributes to social participation. There is a link between participation in education and social engagement as well as between learning and political participation. Older people who continue their education are more likely to volunteer, have more confidence in political institutions, and are more likely to participate in signature campaigns and political discussions, an important aspect of living democracy. Education and a healthy lifestyle keep you fit. How healthy men and women live depends strongly on income and education everywhere. People with a low level of education are much less likely to exercise and more likely to suffer from weight problems than their peers with a higher level of education. Education would therefore also be part of preventive health policy.

If we now intend, in the sense of a good quality of life, to create the conditions for such positive interrelationships for as many people as possible, it is clear that quite different organisational and political areas must come together. Let me remind you once again of the problem of the highly complex division of labour that I mentioned earlier. Education and health policy, family and employment policy, adult education, workplace health promotion, but also voluntary organisations are called upon to work together in this context. Different strategies are necessary: the activities must be developed across generations, across ministries and overlapping organisations. This, it seems to me, is lacking to a considerable extent. The consequences are not insignificant, because we know, for example: Measures taken in the interests of only certain groups can have a negative impact on others, disadvantages that occur in earlier life are exacerbated in old age, and long-term consequences of measures must always be considered in advance, although this is not done regularly. I know that this is not easy; the human capacity to think and judge is limited and perspectival – but there are just enough reasons to nevertheless and repeatedly try what is aptly called "thinking outside the box".

What are the background questions in such more complex empirical contexts, how are they to be connected to far more general considerations? What do we want to understand by education in old age, does the interest revolve around education in old age, for old age or for learning to grow older? In educational practice with elders, these distinctions do not always matter. There, education is provided in a context that is strongly oriented towards the certainty of practical experience in the sense of actively adapting to changing life circumstances and coping with emerging problems in the social world. Not to be forgotten is also the dimension that understands education and the acquisition of new knowledge to deepen self-understanding and to broaden discovery of the world. In contrast to the concept of learning, the concept of education is associated with the idea of a reflexive appropriation of the world and the creation of meaning. Among the assumptions long known in social gerontology, therefore, is that education can be counted on to provide cognitive and emotional support in processing and anticipating changes in mental, physical, emotional and social abilities and events. Cornelia Kricheldorff (2010) has cited as determinants of education in old age: the unfolding of identity and the confrontation with age-specific developmental tasks in a concrete-historical culture and society. She, like many others who have long resisted an economically functionalized concept of education, advocates the use of a holistic concept of education that is valid for the life course throughout the entire age phase. The most important orientations here are: competences for coping with everyday life, action and social competences, but also creative competences. Concerns can be: Self-reflexivity, (self-)experience and (self-)expression. The reflection of life-historical experiences, as an important orientation aid for the conscious shaping of further life, also includes the decision as to which learning and life goals are to be realised in old age.

If we understand education in old age as a cognitive activity that has a strong experiential and reflective component and as a social activity that aims at active citizenship, then we put the intentionality of education in the foreground. Intentionality does not only mean purposefulness, but also meaningful action. There is a great deal of empirical research evidence for the positive effect of meaningful action, which goes far back into early gerontological activity research. However, it remains open whether learning meaningful content is actually superior to learning meaningless content in educational processes. Doesn't precisely undirected learning create openness and opportunities for creative action? Do not societies of the "long life" in particular need learning interests that build on longer and uncertain processes (cf. Kolland 2016)?

Participation

Social participation and social support are closely linked to health and well-being across the life course and in turn clearly demonstrate the link with education. Participation in leisure, social, cultural and religious activities in the community and in the family enable ageing people to show and use their competence, to enjoy respect and esteem and, where appropriate, to maintain or build support and caring relationships. It is extremely important for older people to experience self-efficacy in this way. Overall life satisfaction is significantly higher among those who are socially engaged and engage in activities defined as beneficial to others. As expected, a statistically significant correlation is found with educational status: the higher the level of education completed, the higher the life satisfaction and the less often the feeling of no longer belonging, the more frequent is participation in voluntary activity. The increasingly high level of education of the next generation of older people will therefore play an important role. Especially among women, who make up the majority of the respective older generation (two thirds of the over 70-year-olds are women), a rapid catching-up process can be observed in this respect. Education determines a number of other behaviours and attitudes. In particular, higher education is associated with a more differentiated experience, more varied interests, more extensive social contacts, including outside the family, a higher level of activity and, most importantly, better health, especially psychosomatic health. Older people with a higher level of education are more stimulating, more self-confident; they are much more anxious to maintain their present circle of interests and plan more strongly for the future, into which they also do not submit fatalistically, but which they want to shape themselves. Above all, their non-family activities are more pronounced. This also results in a higher assertiveness, conflict ability and willingness in case of a collision of interests.

I would now like to mention a central idea: education, participation and mobility have a broadband effect on the quality of life in old age; they cover a number of dimensions either as causal or as intervening variables and show a dynamic of cumulative influences over the course of life. Again, the idea of strategies is that they should be intergenerational, interagency, and overlapping organizations. Take social participation as an example, but it could be others. Here it is noticeable that, for example, in social space planning, technical and physical planning are in the foreground, that there is a certain weighting in the interest of youth work, but hardly any for work with the elderly; mobility is understood heavily as a transport matter, but hardly as a problem of social infrastructure; cooperation between educational institutions reveals a lack of dismantling of access barriers in adult education; health promotion is fortunately slowly gaining a foothold in workplace health promotion, but there is a lack of a concerted plan of preventive health care in the preliminary stage of old age (ca. 55 onwards); in the development of new settlements, especially in the area of industrial and commercial settlements, social infrastructure considerations play a minor role; in the offers for voluntary work by older people, too little attention is paid to the fact that this is most likely to be activated when people can organise their time according to their own ideas, the activity is meaningful, and the commitment can be directly linked to the skills they have acquired in life. This is a broad field in which practical work and political measures only capture parts of possible knowledge, i.e. networked knowledge is not used. I conclude these reflections with a note that is important for the whole design.

There is a principle of scientific theory which says: the more complex the content of a problem, the more complex are the logical relations. It pays to always keep this complexity in mind. In another way, Otto Neurath already said this almost 90 years ago: We not only need theories that describe things, but also theories that describe our knowledge.

Here again, background considerations can be used to embed what has just been outlined. In the evaluation of age, the contradictory nature of society itself becomes apparent. Those who carry on as they did in their prime are accused of not being capable of maturity and letting go. Those who moderate themselves in old age get the reputation of becoming senile. Here classical psychoanalysis has not served us well. It persisted in the idea that the human being was formed in early youth. What comes afterwards, up to old age, are modifications on ever narrower paths. In doing so, she completely overlooked what is now slowly beginning to be taken for granted: the plasticity and changeability of the human being in old age. From a very practical point of view, the question may arise here: What to do? A provisional answer may be attempted, prescriptions can probably not be given, but helpful orientations are possible. They can be derived from experience, from sociological and psychological research on ageing, from philosophy and literature. Two keywords are to be mentioned: Competence and self-attention.

Competencies are those abilities that lead to a way of life adaptable to changing conditions. Certainly the following belong to it: Social relationships in which social recognition is experienced, avoidance of withdrawal and unbalanced exchange. Those who always give and give and give will inevitably become dissatisfied. To be able to deal successfully with new tasks, to be able to learn from conflicts, to be able to respond to one's partner, even in difficult situations, is an expression of competence. Women and men who succeed in their partnership in developing new goals and harmonious coexistence in a new situation after retiring from working life thus demonstrate a very high mental and spiritual capacity. Competent feel those old people who are recognized because they fill important and meaningful roles, who can control their environment, have an appropriate self-expression and are open-minded. Those who feel incompetent are those who are constantly confronted with expectations of error in communication, who are prone to self-disclosure and repetition of the same old, and who thus get on the nerves of others. Many of us remember the men who talked incessantly about the war, which was already far in the past. They were bored by the well-known stories, one did not want to hear them any more. The fact that they often couldn't help it, that they were inadequate attempts to cope with devastating experiences, was seen less often. In the frequently discussed "coming to terms with the past", which is emphatically rejected by so many, this side has been given short shrift altogether. Thinking about one's own ageing requires the development of clear ideas about one's self, and these are shaped through social relationships and one's own behavior. Without figuring out the deeper reasons for our behavior, it never succeeds.

Self-attention means keeping an eye on the course of one's life and avoiding incidental, non-purposeful developments – "life management" is a good word for it. "It just happened that way," can often be heard, especially in bad situations, and "I should have done things differently" is the belated insight. Especially in the face of a long life, "living it up" becomes nonsensical. This has nothing to do with constant self-control, or with uncreative, boring rule-following. The correct assessment of one's own abilities and qualities has a decisive influence. Thinking about one's own goals and wishes, one's own possibilities and limits is important here. Successful ageing is closely related to mentally anticipated plans and activities. This anticipation is an important prerequisite. Here the way could be found to a discussion about the old age personality. Age, and especially old age, is also a phase of losses and gains, of upheavals and new beginnings. The willingness to follow these changes productively depends on how much the respective person has a sense of self-competence and self-efficacy, how much attention is paid to the self, how the self-assessment compares to reference groups. For many, the death or serious illness of a loved one is a catastrophe that is almost impossible to cope with. Most of us have lost the ability to accept such strokes of fate out of deep, religious faith. All the more we are challenged to deal with it out of insights into our own lives and their changeability (Amann 2004).

There is the beautiful story by the German writer Barbara König, "Waiting for Wisdom". A woman, good-looking and active in life, has two drastic experiences shortly before she turns 60. A blond-haired youth at the train counter asks her for a "senior citizen's pass" when she goes to buy a ticket. She experiences a shock and begins to wonder if and when old age begins. From her own life, the relativizations offer themselves. "At eighteen I loved an old man of thirty-six; on my fifty-nineth I danced with a youth of forty-five. On my eightieth I could, theoretically, flirt with a peer of seventy." The second assault on her self-image of her own age comes the next morning at the hotel, when she is packing her bags. She can't find the pin with the two sapphires. Suspicion shoots out of her on all sides, mixed with presumed forgetfulness of age and readiness to accuse. She is ready to rip the mask off the face of anyone around her. Then the pin is found on the costume she wore the previous evening. But a trace of mistrust remains (Amann 2004).

For the time being, no help can be expected from society for such experiences. It is self-contradictory in its judgments. Deep underground, old moral ideas of obligation and responsibility are mixed with the principle of utility and the demand for the usability of human life for material purposes. Capitalist production can exist only on the exploitation of human labor power. The "meaning" of a human life ends where this exploitability ends. Rigorously thought further, this means that for economic production the old who (can) no longer be employed are in fact useless and obstructive. Of no value is that from which no capital can be made. Fortunately for them, some can still consume vigorously. Thus, in the eyes of the rigorous representatives of this ideology, state old-age provision is also considered antiquated. Old age consistently begins with the impossibility of selling one's labour power. The approach to overcoming this must therefore be sought elsewhere. It lies in a reluctant movement of individuals against their own declaration of uselessness. The destroyers of meaning in old age must be taught by the bitterest possible experience that they have their existence solely through those they disregard. But this is a project that does not begin with today's 60-year-olds. Those who are 20 today, and even those who have not yet been born, are already potentially subject to the annihilation of the meaning of their age. The long breath of the age sense destroyers reaches across the generations. The unseen perfidy of the demand for private old-age provision means, after all, that one is supposed to provide oneself for a situation that will be imposed on one by society. In Otto v. Bismarck's day, the age limit and invalidity still coincided. In order to find a way out of this, the state pension scheme was invented. Today the age limit and the impossibility of selling one's labour power again (still) coincide, but the solution is to pass on the worry to the individual. The rudder to steer by is the education that society is more than economic production. There is an indissoluble connection between economics, culture and society in which old age is embedded. Not to see this connection, to simply separate old age from it, is to lose sight of the social basis of existence (cf. Amann 2004).

7.2.2 Regimes and Perspectives in Life

In Sect. 3.2.1 I have already referred to the importance of the life course perspective, for a consideration of participation in ageing can only gain from this conception. We can assume with some justification that a scientific consideration of the life course has only begun to develop in sociology since the 1970s. Since that time, both the reconstruction of individual biographies and the reconstruction of collective life courses have been used to analyse the social structure of society. The welfare state and social policy have made a strong contribution to the shaping of life courses as well as to societal change as a whole; in particular, since the 1990s many studies have looked at the pension system as a structural generator. (e.g. Leibfried et al. 1995, p. 23). Thus, in addition to the previously well-researched structure generator, education or the educational pathway, a second one had been established, which was used to describe structured status transitions (to work, to retirement). This idea of formation, which at the same time gives the life course the character of continuous change, is nourished by two moments, one of which I have already described as external, the other as internal resources, and which at the same time coincide with the time dimensions of life situations. Martin Kohli (1978) distinguished between "biography" as a rather subjectively conceived life history and "life course" as an objective event history. Later, Karl U. Mayer (1990) and Helga Krüger (2010) should be mentioned for a critique or shift in emphasis and a reshaping of the terminology, and Wolfgang Clemens (2010) for a concise overview of the research tradition on the life course.

What insights can be drawn from these conceptions? Now I want to take account of the last term of the chapter title at the end and, using the example of the life course, look more closely at various structures that are connected with the topic of participation. The consideration should be broad, as the following examples show. Life courses are generally influenced by demographics, in that they change life expectancy and shape the proportions of people in different age groups; they are also affected by sectoral policy decisions and legislation, for example when changes in the education system (duration of schooling and studies, international exchanges, etc.) take effect. Above all, there are social policy and social law programmes and provisions that take effect, for example, through working time regulations and pension age limits; significant changes in the world of work play a role in that different forms of employment, levels of pay or unemployment become effective; finally, there are changes that are due to general social change, such as postponement of the age of marriage, changes in the age at first birth and the average number of children, migration, etc. (cf. also Naegele 2010). In order to bring a heuristic order to these forms of impact, I will stick, with some modifications, to a classification proposed by Gerhard Naegele (Naegele 2010), because it can be well approximated to my own conception of social policy and change in standards of living over the life course (Amann 1983).

From the point of view of social policy, attention should be focused on risks that are effective or newly emerge in the course of structural changes; social policy thus has a preventive and a compensatory shaping task (function of shaping society by intervening in standards of living). Such risks occur both at different stages of life and over the course of life. In order to bundle these risks thematically, a division into (1) standard employee risks and (2) general standard life risks has been established for some time. In the first case, for example, problems of young people entering working life or employment problems of older workers would come to mind. In the second case, the risk of single parenthood or the risk of needing long-term care in old age would come into consideration. This would be distinct from special risks that arise in the event of changes in the law, such as in Austria with the new minimum security regulation since 2018. This classification is not necessarily compelling from a factual point of view, but can be retained due to its prevalence.

It is clear from this observation that all effects have a considerable direct or indirect influence on, or even determine, people's participation in life. Again, empirically proven correlations can be presented to demonstrate this.[7] Disadvantages in the choice of occupation impair later career development prospects; previous unemployment reduces future income opportunities, with increasing significance in recent decades; low-income earners have an increased risk of becoming disabled at an early age; precarious employment and long-term unemployment increase the risk of permanent economic underprovision and poverty in old age; women's care work for children impairs their later career opportunities (and indirectly also their pension income); single parents (especially mothers) have lower income opportunities and consequently less favourable pension expectations; overall, part-time work (mostly for women) reduces later career and income opportunities as well as their own pension entitlements; health and illness in old age can usually be understood to a large extent as the result of a cumulative positive or negative development of life circumstances.

Since some age-related empirical correlations that can be counted under the heading of standard life course, albeit under mostly negative auspices, have already been dealt with in detail above, it may suffice here to highlight a few more general considerations on the life course issue in connection with age. One of the most socially significant effects of demographic change that affects ageing is probably

---

[7] For the individual sources, see Gerhard Naegele (2010).

the ageing of the working population. The broad trend for many countries is a decline in the size of the labour force combined with internal ageing. A long-standing conviction is that the working world of tomorrow and the day after will have to be managed by fewer and older workers overall (Amann 2004). However, as a cohort effect, the health and qualifications of future older workers are likely to improve. Lifelong learning and employability are also likely to experience a positive alternating effect, which will be of particular interest if the actual age at which people retire increases noticeably. Lifelong learning is associated with higher productivity potentials that are supposed to extend far beyond the statutory age limit, and an explicit link is established with the life course insofar as educational obstacles and educational opportunities that become effective in later phases of life are in turn an expression of previous educational careers (Naegele 2010, p. 52). In view of the manifold references, the call for a social policy that takes greater account of life courses is a democratically sensible demand.

## 7.3	Signs of Significant Change

The "New York Times International Weekly", the Monday supplement of the Austrian daily newspaper "Der Standard" (01 April 2019), featured the following story by Nellie Bowles. Bill Langlois, a low-income retiree, lives in "a low-income senior housing complex in Lowell, Massachusetts," and he is lonely. But, he has made a new best friend: She's a cat named Sox and lives in a tablet; she makes him so happy that tears come to his eyes when he talks about her; in the past, it would have been said that he occupied this object libidinously, which at least describes the heart of the matter well. Sox and Mr. Langlois chat all day, which means they communicate in real time. Sox plays him his favorite songs, shows him pictures from his wedding, and reprimands him when he drinks soda instead of water (the tablet films his behavior and responds according to the parameters installed in the background). Of course, the man knows that Sox is an artificial being and comes from a start-up called Care.Coach, which is worked on by people around the world "who are whatching, listening and typing out her responses, which sound slow and robotic." Mr. Langlois says he found something so reliable and so caring for him that he found his way back to his faith in God. "She's brought my life back to life." Sox, of course, overhears and replies, "We make a great team."

There is no need to reproduce more from the story, the message is clear enough. Here, a dimension that I have so far only touched on in passing, that of the information-technological penetration of the social world, or, to put it another way, creeps into the discussion of participation: Life is increasingly mediated to more and more people via screens. In the context of the concept of participation in life presented here, this means nothing other than that in recent decades people have constructively formed a new strategy or function within the framework of their ways of life that allows them to act, speak, and think about each other in real time without face-to-face contact. Programs, like the one mentioned above, are proliferating, and it's not just the elderly who benefit from them, but almost everyone, especially those who can make money from them. Any place, from the classroom to the hospital to the airport to the church to the football stadium, from the bedroom to the car, where a screen can be usefully installed can help save money and increase curiosity about the world. This development seems to have meant that people are no longer "at home" where they "live". The impression is difficult to ward off that people are being drawn away, daily, hourly, year in and year out, as if in a maelstrom, into foreign, alluring, provocative, sometimes even instructive world districts. At the same time, the cause and effect of this development is the incessant replacement of the new by the newest and of the latter by the very newest. Now, this incessant and ever more rapid change from the new to the newest does not necessarily have to be opposed by the traditional. That would be to misunderstand the logic of the constructive production of life forms by human beings. What should be at stake, however, is preservation in the literal sense, preservation in contrast to the unreflected squandering and destruction of nature and of what has been historically created for the sake of short-term gain. It is hard to imagine, but it is possible, that the children of the next generation but one, or the one after that, will be shown films of rainforests or intact small mountain villages in school to illustrate something that no longer exists. If we no longer know any place where we feel permanently at home, even in thought of it, then, one suggestion might be, the other places that excite our curiosity no longer have any value either. If the idea of a home, which can rightly be called such, disappears from the world of today's people, we should ask what other things are being prepared in the midst of the rush of this brave new world. Should hope be placed in this other, it will be disappointed at present. Rather, the return to the before is practiced again, but it can no longer be a return; no one climbs into the same river twice. Thus, even from the numerous individual pieces of information about allegedly new Internet behavior (rich people send their children to Waldorf schools without Internet equipment, young people give themselves a text message-free day or do not hang out on the smartphone for a day and evaluate this as behavior that signifies special status), no new quality emerges yet that could dispel the concerns that are carried along everywhere. Participation mediated by information technology is an essential dimension of participation in life at the current level of development of the forms of life we have created; what long-term effects it produces, whether for the better or for the worse, remains to be seen.

Structural Disadvantages and Impeded Participation

8

As soon as a compound is used in a relaxed conversation, in which the word "social" occurs, many people reflexively cry out: "We are far too social" – whereby unspoken mostly asylum seekers, so-called social parasites and other notoriously stigmatized groups, minimum security recipients and often also the wealthy old people are meant, "who travel around the world", and who therefore have too high pensions. On such occasions, the cabaret artists of public opinion (Helmuth Plessner) inevitably get the upper hand. With a high degree of reliability, such reactions can be expected from the supporters of right-wing populist parties and politicians (including, of course, the politicians themselves), but also from the unthinking. Without searching too extensively for reasons for such attitudes, it is probably true that they are based on false information, which they themselves spread in the social media, but also in tabloids almost like thunder. "I am of the opinion that the foreigners get much more than the nationals, if you don't see it that way, that's just your opinion. I see it that way."[1] It's no longer about facts, everything is dismissed as opinion and yours is valid no matter what. Now it is indeed not always easy to find balanced figures, and the process of reliably informing oneself takes some effort and practice. Remembering the discussion on the standard of livig concept, it was noted there that the first scope mentioned by Ingeborg Nahnsen was the "supply and income margin (extent of possible supply of goods and services)". This is the one to start with here. In comparison to the misinformation mentioned above, it is precisely here that well-prepared and balanced accounts can be found, as can be seen in the example of the risk of poverty, reliable evidence of a narrowed life.

---

[1] Verbatim quote from an inn conversation.

Hardly any empirical finding is as clear as the decrease in social participation with low or too low income – even in connection with weak social networks and compromised health.

8.1   Gender-Differentiated Ageing Trajectories in Germany[2]

Similar to comparable countries, more and more people in Germany can expect to live longer and longer. Above all, this is expressed in the average life expectancy at the time of birth, which is now 78,6 years for men and 83,4 years for women (2019/2021). The question at the centre of this report by the Federal Ministry for Family Affairs, Senior Citizens, Women and Youth is whether life courses differ between men and women, in particular whether they differ in the second half of life, i.e. from middle to very old adulthood, and which characteristics can be used to observe these differences. These characteristics thus refer to gender-specific ageing trajectories and their change over successive birth cohorts in various life domains such as health, life satisfaction and depressive symptoms, loneliness and social isolation, caring activities as well as voluntary engagement (I will single out a few). The general answer is: "Life situations in middle and old age are characterized by manifold gender differences: Women are more likely to have better social integration than men and they take on caring activities to a greater extent than men. Older women live alone more often than older men. Women also suffer more frequently from depressive symptoms and, in older age, more severely from declines in functional health than men" (Bundesministerium 2019, p. 6). These are results that were also found for Austria.

In social gerontological research, subjectively assessed health is regarded as an important factor influencing the experienced state of well-being, the perceived quality of life and the level of social activity. It is well known that with increasing age, women and men assess their health less positively. However, according to the empirical data, the subjective assessment of one's own health over the course of ageing from 40 to 90 years decreases less strongly overall than the ageing course of functional health would suggest. It is interesting that women and men subjectively assess their health similarly over the entire second half of life, although women report more impaired functional health than men and this becomes worse for them with increasing age than for men. This is a finding that, for methodologi-

---

[2] The statistical figures and some interpretations are taken from: Federal Ministry for Family Affairs, Senior Citizens, Women and Youth (BMFSFJ) (2019).

cal reasons, makes it almost imperative for empirical research to always observe both operationalizations (subjective assessment and functional status) simultaneously in order to prevent one-sided assumptions. Not surprisingly, there are no cohort differences for subjective health: women and men of later birth cohorts do not differ in their health assessment either in their initial level in their early 40s or in the course of ageing (Bundesministerium 2019, p. 12).

The next area to be considered is life satisfaction. In some operationalizations of quality of life it is one dimension among several, here it is considered alone. It decreases somewhat at an older age and is different for women and men. "Around the age of 60, women are on average initially more satisfied with their lives than men; however, with increasing age, a decline in life satisfaction sets in, which is more pronounced for women than for men" (Bundesministerium 2019, p. 14). This result also confirms insights from other countries, so with some caution it can be considered stable in cross-sectional country comparisons, but not over time, as subsequent cohorts show more favorable ageing trajectories of life satisfaction, with no differences between women and men. The decline in life satisfaction with increasing age is more pronounced in the earlier born cohort (born 1930–1939) compared to later born cohorts (born 1940–1949 as well as 1950–1959). Obviously, these findings are closely correlated with the next area. Depression risk increases with age, and clinically prominent depressive symptoms are generally more common in women than in men. The likelihood of the occurrence of these symptoms also increases more with age in women than in men. Subsequent cohorts do not differ significantly in their ageing trajectories with respect to depressive symptoms; this is equally true for women and men. The difference in the ageing trajectory between women and men is therefore similar in all birth cohorts (Bundesministerium 2019, p. 16).

Since isolation and loneliness are repeatedly highlighted as significant categories for participation in life in this book, attention should also be focused on this once again. The data from Germany provide a familiar picture. It goes without saying that "social isolation" and "loneliness" are two different concepts (even if too little precise distinction is often made between them). Social isolation is measured by a comparative lack of contact with other people; it is observable. Loneliness, on the other hand, is a concept meant to describe a psychosocial subjective experience. Now, the opinion is indeed very widespread that in old age the risk of being socially isolated and/or lonely is enormously high, and since the number of very old people is constantly growing, the cases of isolation and loneliness must also increase. The real finding is that the risk of social isolation increases with age, and it differs between women and men as they age. "For men, the risk of social isolation increases relatively uniformly from five to 20 percent over the age range considered, from 40 to 90. Women initially experience a weaker increase in risk, but this accelerates in retirement age, so that by the time they reach their late 70s, they are similarly likely to be socially isolated as men. Prior to this, women have a lower risk of isolation than men for more than three decades of life from age in their early 40s to mid-70s" (Federal Ministry 2019, p. 20). Moreover, the risk of becoming isolated as one ages has changed in recent years. For later-born cohorts, the risk of isolation no longer increases to as high a level as for earlier-born cohorts as they get older, and for all cohorts, similar gender differences in the ageing trajectories of this risk are evident. This may be taken as an indication that exogenous resources have changed for the succeeding cohorts, but would require a multilevel analysis.

The risk of loneliness differs between women and men as they age in that in middle adulthood, between the ages of 40 and 60, men are slightly more likely to feel lonely than women. "However, the gender difference decreases with age and reverses during retirement, so that more women than men are lonely at older ages. At age 90, women have a 14 percent risk of being lonely. For men at this age, the risk of loneliness is nine percent" (Federal Ministry 2019, p. 22). Data also suggest that loneliness risk is less strongly associated with ageing in later-born cohorts. "In terms of loneliness risk, the data show a lower baseline level in middle adulthood for those born between 1950 and 1959 and a flattening of the U-shaped aging trajectory. Thus, it is likely that those aged 70 to 80 in the future will be less likely to be lonely than those aged 70 to 80 today" (Federal Ministry 2019, p. 22). Furthermore, there has been no discernible levelling out of the ageing patterns between women and men to date: similar gender differences in the ageing patterns of the risk of loneliness are evident in all the cohorts considered.

The results relevant to participation are summarised as follows: Women have a more limited functional health than men throughout the second half of life. The difference between women and men increases with age; there is no statistically significant gender difference in the subjective health assessment, neither in the initial level nor in the course of ageing; women are more satisfied with their lives than men in middle adulthood, but more dissatisfied than men with increasing age; Women have a higher risk of depression, which also rises more sharply than men; in return, they have a lower risk of isolation than men until older age (up to about age 80), and a higher risk thereafter; women have a lower risk of loneliness than men until the seventh decade of life, and a higher risk thereafter (Bundesministerium 2019, p. 36).

8.2   Poverty Risks in Europe[3]

At-risk-of-poverty rates are an internationally equivalized measure that allows comparisons between individual countries; it denotes an extremely limited scope for monetary action. The data presented below are based on the results of the EU-SILC (EU Statistics on Income and Living Conditions) from the statistical office of the European Commission (Eurostat), collected in 2016 and published by Eurostat in 2017. The EU-SILC is an annually repeated survey in all EU countries and serves as a reference source for comparative statistics on income distribution and social inclusion in the European Union. The at-risk-of-poverty rate indicates the proportion of people at risk of poverty in a total group. People are considered to be at risk of poverty if their income is less than 60% of the median income. The income calculation takes into account both the different household structures and the savings effects that arise from living together – through shared living space, in energy consumption per capita or in household purchases. Thus, incomes are weighted. The at-risk-of-poverty rate is measured here in relation to the situation in each country, rather than using a uniform threshold for all countries. It is therefore a relatively differentiated measure that provides far more reliable information than the usual, usually misleading, data on average pension incomes. Household disposable income is the sum of the total income of all household members from all sources (including income from employment, investments and social benefits), with income added at the household level and taxes and social contributions deducted. To take account of different household sizes and household compositions, the total is divided by the number of "adult equivalents" using a standard (equivalence) scale. This "modified OECD equivalence scale" weights the first adult living in the household by 1.0, all other household members aged 14 and over by 0.5, and household members under 14 by 0.3. The equivalized income calculated in this way is allocated to individual household members. For the compilation of the poverty indicators, the equivalized disposable income is calculated by dividing the total household disposable income by the household equivalized size. Consequently, the equivalized income is the same for each person living in the household.

The following at-risk-of-poverty rates were collected once before and once after social benefits received in selected EU countries in 2016. Now, it is well known that as a means to reduce poverty, all countries use different social benefits. One way of assessing the success of social protection measures is to compare poverty

---

[3] https://de.statista.com/statistik/daten/studie/1171/umfrage/armutsgefaehrdungsquote-in-europa/ (retrieved:03/18/2019). The percentages were rounded, partly verbatim reproduction.

risk indicators before and after social transfers. The five countries with the highest poverty risk after social transfers in the EU in 2016 included Romania (25.3%), Bulgaria (22.9%), Spain (22.3%), Greece (21.2%) and Italy (20.8%). These are exclusively countries from Southern and Eastern Europe. The economic and financial crisis can only be used to a limited extent as an explanation for the relatively high poverty risk, as the values were already consistently high before the onset of the crisis in 2007. In contrast, lower poverty rates below the EU average are predominantly found in central and northern Europe, such as France (13.6%) and Sweden (16.2%). If the redistributive effect of social benefits is not taken into account (with the exception of old-age security), the at-risk-of-poverty rate in the EU countries increases, sometimes substantially. The extent to which social benefits reduce the risk of poverty varies between countries. If the countries are sorted according to the level of poverty risk rates before the redistribution of social benefits, the result is a different order than after the redistribution. Without social benefits, the risk of poverty in 2016 was highest in Sweden (29.9%), Spain (29.5%), Romania (29.5%), the United Kingdom (28.1%) and Bulgaria (27.9%). The most significant reductions in the risk of poverty among the population in 2016 were achieved in Sweden and the United Kingdom, where the incomes of about half of the population at risk of poverty were raised above the poverty threshold. Furthermore, the risk of poverty in France (−42.4%) was also greatly reduced by social benefits. In Germany, the reduction in the population at risk of poverty was 34.8%. In relative terms, social benefits in Greece (−15.9%), Romania (−14.2%), Bulgaria (−17.9%) and Italy (−20.3%) brought about the smallest reduction in poverty among the respective populations. Thus, the social systems in Central and Northern Europe have a supposedly greater impact than the social systems in Southern and Eastern Europe.

The background is simple to state. The fight against poverty is one of the most important social policy objectives of the European Union. Nevertheless, in 2016, around 87 million people in the EU lived in income conditions associated with a risk of poverty. To illustrate this fact, it is sufficient to imagine that the populations of Germany and Austria together would be at risk of poverty. A person is at risk of poverty if the household income available to each member of the household is not sufficient to purchase the goods and services needed to cover the socio-cultural subsistence minimum. The national at-risk-of-poverty thresholds vary widely in this respect While the threshold in Germany in 2015 was EUR 1064 per month for a single person, in Spain the amount below EUR 684 per month was already considered to be at risk of poverty and in Bulgaria below EUR 158. In this context, one therefore speaks of relative poverty. Different groups in society are at risk of poverty to different degrees. There is hardly any distinguishing characteristic that has a greater influence on the extent of poverty risk than occupational status. In 2016, the overall at-risk-of-poverty rate for the unemployed in the EU was 49%, more than five times higher than that for the employed at 9.6%. In no EU Member State was the at-risk-of-poverty rate for the unemployed higher than in Germany (70.8%). In 2005, the at-risk-of-poverty rate for the unemployed in Germany was 40.6%, in line with the EU average. This reflects the impact of labour market reforms in Germany. In particular, the introduction of the so-called Hartz IV Act as a basic security for the unemployed is increasingly not sufficient to finance a living wage. Furthermore, an examination of different household types in the EU for 2016 reveals a particularly high risk of poverty among persons living alone (25.6%) and single parents (33.8%). In addition, the risk of poverty is highly correlated with the level of education acquired, age and gender.

8.3   Risk of Poverty in Old Age in Austria

As already noted, Austria is not included in the above-mentioned overview, but the patterns observed can also be found in this country. The focus will now be on the situation in old age. Unfortunately, no recent special evaluation is available at the present time (March 2019), so that it was necessary to fall back on a publication of the BMASK from 2012 (Eiffe et al. 2012). The social security in old age (the age limit is defined as 60 years), i.e. the material prerequisite for participation in life, is for 81% of older people their pension benefits. But also non-pensioners living in the same household are often dependent on these pension benefits. The importance of this form of income in old age is most clearly seen in the fact that 78% of people over 60 would fall below the at-risk-of-poverty threshold without pension and social security benefits. "Receiving pension and social security benefits reduces the at-risk-of-poverty rate for this group to 14%" (Eiffe et al. 2012, p. 15). Who is actually poor in old age is determined by occupational position. Women are most affected by poverty risk in old age, persons of higher age, single female pensioners as well as persons without employment involvement or with a low-skilled occupational position at an earlier working age. Women are 16% at risk of poverty, men 11%, and the risk increases with age in both groups. A bridge can be built from here to participation issues. Social isolation affects people in old age more frequently than the population on average. In Austria, at the time of the survey, there were 215,000 people aged 60 and over who had no regular contact with relatives, friends or neighbours. In any case, the statistical analyses clearly show that social isolation is not a fate in old age, but is closely related to restrictive living conditions (Eiffe et al. 2012, p. 19). We all know those people who give the impression of having renounced company at some point and sought a place to which they could retreat for good; they do not take an interest, do not lose a word and let others pass by like the seasons. This brings up for discussion not only the unequal distribution of material resources, but also the strength and scope of social networks and cultural integration. But it is precisely these networks that are an important prerequisite for coping with financial or social hardship.

The primary social network in which social integration takes place in contemporary societies is the multi-person household. With age, however, the proportion of people living alone increases, especially that of women living alone, who can no longer fall back on this primary network. It is not uncommon then for an older woman to spend 20 or more years alone in an apartment or house after the death of her husband and the departure of her last child. Almost every second woman between 70 and 79 lives alone, but only every sixth man in this age group. At older ages over 80, the majority of women (54%) live in one-person households, but less than a third of men (25%) (Eiffe et al. 2012, p. 114). In addition to family ties, there are often contacts with relatives, friends and neighbours. If age groups are contrasted with one another, it is true for men, and less clearly for women, that a change in the three fields of contact mentioned is visible. In younger working age, regular contact with friends is more frequent than with relatives and neighbours. However, the picture changes over time. Contacts with friends decrease significantly with increasing age, but the proportion of people who have regular contact with relatives and neighbours remains constant. Women are more likely to have regular contact with relatives than men in all age groups, and contacts in the neighbourhood become more important from middle age onwards. The contact patterns that occur between 60 and 69, with no difference between men and women, hardly differ from those in later working age. Among those aged 70 to 79 and those aged 80 and over, the frequency of contact with friends then falls for both women and men, and contact with relatives declines slightly but remains significantly higher for women (Eiffe et al. 2012, pp. 115 and 116). If people have no regular contact with friends, relatives or neighbours, they are considered socially isolated. Indeed, the proportion of men and women who are socially isolated increases with age, but in each age cohort men are more likely to be affected than women. The further prevalence of isolation at retirement age is due to the loss of involvement in working life. Among the over-80s, 17% of women and 19% of men are isolated (55,000 people combined). "Overall, around 103,000 women and 111,000 men over the age of 60 live in social isolation" (Eiffe et al. 2012, p. 117).

What about the so-called extremes of expression in these analyses? First and foremost, and most clearly, it becomes apparent that the risk of poverty among persons living alone in the 60+ age group is predominantly female. 84% of the one-person households living below the at-risk-of-poverty threshold are women. Further, the group at risk of poverty suffers from poor health more often than the general population. People aged 60 and over who are not at risk of poverty report very good or good health in 45%, compared to 23% among people of the same age who are at risk of poverty (Eiffe et al. 2012, p. 167). Falling out of social networks, disengagement from participation, has been mentioned several times. With a little effort, knowledge about resignation, withdrawal, experiences of loneliness, about failed attempts at contact and longing for a fuller life among these people could also be found in the psychology of old age. Obviously, such analyses provide enough material to counteract or correct the prejudices mentioned at the beginning. It would be a worthwhile research question for adult education why people who are now of active age do not and do not want to learn that all the cutbacks in European social systems that we have been experiencing for more than 20 years, and which are intensifying under the current right-wing populist governments, will probably still affect them themselves, but certainly their children, and why they nevertheless continue to support these governments.

In this chapter, structures are in the foreground of attention, the last topic dealt with was disadvantage, and following diagnoses, suggestions for practice can be expected. In the following, I will summarise a number of considerations that are well known in a wide variety of publications from sociology and social and economic policy; detailed references are unnecessary in the case of such a clear presence. Which strategies for combating material deprivation and its consequences are relevant in practice? From the point of view of de facto deprivation in old age, traceable to employment histories, the following are the most obvious: a re-regulation of employment relationships and wage structures, a continuous reduction of unemployment, especially its long-term form, as well as stable reintegration after loss of employment, the enabling of continuous employment for women, the improvement of the reconciliation of work, child rearing and family care, as well as the equalisation of female and male wage and salary levels. These strategies are seen as preventive, but they are not a panacea. At least for Austria, a policy will be necessary in the future that does not allow the following correlation to take effect in the first place: If the wage (especially in the case of part-time employment) hardly secures the individual subsistence level and the livelihood can only be guaranteed in the context of a partner, it is certainly not possible to expect pensions that are higher than the minimum security or social assistance. It is to be seen as a politically and economically motivated attack on the future social security of entire population groups that a problem preceding the aforementioned issue is simply accepted and even promoted by tax and social law regulations: the expansion of part-time work, especially on the basis of mini-jobs. Although all those who are interested in this issue have long been aware that a massive problem for the future is being discussed here, most of them appear to be unconcerned – incomprehensibly so in view of the development. Assuming that the old-age security system remains roughly in its present form, the picture is clear. The future income of the elderly, their pensions, will be determined in terms of level and distribution by a whole series of economic, socio-cultural and political-legal factors which, on the whole, cannot be predicted with sufficient accuracy. This is why multiple preventive strategies are so important. In the socio-political discussion, a distinction is made here between exogenous and endogenous factors. The latter refer to the foreseeable cuts in benefits and regulatory changes in old-age insurance systems, especially pension insurance. In the case of exogenous factors, the question arises as to whether the employment biographies and thus the individual pension entitlements of the cohorts coming up to retirement will develop. Despite the minimal declines in unemployment that have been repeatedly hailed in the media in Austria in recent months, I note, with a view to the exogenous factors, that a veritable problem and risk potential has been building up on the labour market for about fifteen years (cf. also Bäcker 2016). A few keywords may suffice to describe facts that cannot be eliminated, no matter how much euphemism is used. Unemployment and especially long-term unemployment have decisively shaped the employment biographies of many cohorts in recent decades. Forms of employment that are not subject to compulsory pension insurance have grown, employment histories have become discontinuous, and multiple switches between regular and precarious employment have increased. The expansion of the low-wage sector and of part-time work already show their effects. For the coming cohorts it is therefore to be feared to a considerable extent that pension entitlements will decline. The extent to which unemployment, career breaks, precarious employment, low wages and part-time work have a negative impact on the level of pension entitlements depends not only on the specific pay position but also crucially on the duration (Bäcker 2016, p. 68). All these facts cannot be disputed and can be proven with the help of relevant data. The fact that virtually nothing is being done with regard to the old-age provision of future cohorts is, to put it mildly, a political scandal – if only because this country is still one of the 16 richest in the world,[4] and because at present a shabby lust for fear, narrow-mindedness and raw pleasure in petty nastiness against the socially weakest make up the big narrative, especially on the side of the political right wing including big parts of the traditional Christian-social party, instead of a comprehensive, expertise-based reform discussion of the pension and care system.

---

[4] In the International Monetary Fund ranking, Austria was ranked 16th in 2017 (after dropping four places in two years).

8.4   The Suddenly Narrowed Life

There are thousands of possible interactions between endogenous and exogenous resources of the standard of living that come into question for the individual to shape his or her participation in life, and from which only a part is ever used. What someone may experience and use as learned dispositional scope is closely related to the learning processes from earliest youth, but is also controlled by changing social conditions, the structural moments. The deeper a person's insights into the conditions in which he or she lives, the greater the overall scope that opens up for him or her within the recognized set of conditions. Institutionalized forms of action, habitualized perceptions and decision-making procedures support action in everyday life; the entire context in which participation takes place has a certain pre-reflexive effect through all situations of action. This is the core of the experienced normality of events in everyday life. What has been called, with reference to Johann W. v. Goethe, the "banality of everyday life", shares in this normality. It keeps us in a status of always-so-forth, which relieves us and makes it possible to go about the business of daily participation without being forced to constantly make decisions in situations that are constantly unclear. A person, like most others, who is completely absorbed in his world, does not ask himself the question of the meaning of being, nor does he ask himself the question of the meaning of his own life. Only an experience of loss of meaning, the experience of a relationship to the social world that is disturbed in whatever form, will provoke such questions. The normativity of expectations discussed above (which carries with it validity) and the power potentials or their distribution in the fields of action support this process. On this basis, people hardly ever feel the need to think about possible drastic changes in advance, not even about their own age, which will come one day or is already here. Surveys have shown that even people aged 60 and above only give about 40% of their thoughts to their own ageing, whereby there are, "once again", one might say, considerable differences according to the level of education. The irony is, as the psychology of old age tells us, that it is above all the ageing and illnesses of one's own parents, deaths, and not least one's own health restrictions, i.e. sudden constrictions of the possibilities of participation, which cause people to concern themselves with their own ageing. Now there is a certain tension in the relation always-so-forth/planning-ahead, which is not easy to resolve in everyday life. To constantly dwell on one's own finiteness ("you know neither the day nor the hour"), as demanded in some ancient but also Christian teachings on salvation, seems unhelpful in an enlightened understanding of the world; but to approach a similar inevitability in life, namely old age, with determination would be closer. This is not, however, to draw a parallel with Martin Heidegger's continuous run-up to death, for the question here is not a philosophical justification of being and existence, but a reflection in everyday theory on a problem little considered by the individual. In the above-mentioned sense of a recognized overall scope, there would certainly be the possibility of taking "fate" into one's own hands, namely through the process of an exploratory development of the dynamics that shape one's own life in everyday life. The prerequisite for this would be: recognizing who one is, exploring what one wants, experiencing what one is capable of. These are not detached demands of a philosophical program for esotericists of ideas, but tangible questions that come to everyone again and again, to the point of almost literal formations of concepts in everyday life. "Who am I to be treated in this way?" "What do I actually want for myself in this whole daily circus?" "Am I likely to make it, or more likely not?" Who hasn't had questions like these run through their head? It wants to seem to me that they are part of the everyday repertoire of questioning oneself, even if it may be done rhetorically in many cases.

According to all the empirical evidence, the effects of a suddenly narrowed life in old age are, on the one hand, endogenous and/or exogenous events such as accidents, sudden illnesses, loss of partners, and, on the other hand, the consequences of inadequate planning and foresight, such as long unaccepted losses which then exceed a risk threshold, overestimation and misestimation of oneself in physical exertion, neglect of the shaping of the environment (most accidents among the elderly, for example, occur in the home), and finally financial losses, mainly after retirement. Unquestionably, they limit the possibilities of participation, often to the point of almost total disenfranchisement, as will be found in those groups to which social isolation of the most severe kind must be attested. In this context, it cannot be repeated often enough what has long been known in social gerontology: early occupation with and preparation for one's own ageing helps both to prevent or delay some limitations and to discover new possibilities for successfully compensating for losses (cf. Amann et al. 2010a). Participation in life in old age can be shaped to a high degree, but it is indispensable to take on this new role actively, while at the same time acknowledging that people become frail and that life is finite.

Epilogue

9

It has now been sufficiently clarified with some certainty that participation in life represents a non-reversible moment of the constructive production of human forms of life, anchored in the anthropological constellation of the human being, which makes mutual interdependence a prerequisite for social life in general. If the change of perspective from this epistemological-critical reconstruction of the relationship between man and the social world as well as man and the biological substrate (nature) of life, of which he is also a part, is carried out to a historical-empirical consideration, the space of possible conditions of participation virtually explodes and makes a more comprehensive and at the same time integrating view increasingly necessary. From the perspective of research, however, it seems to me that in recent years, contrary to a comprehensive view that is necessary here, a spray valve effect has come to the fore in the topic of participation. One could almost be taken aback by the way in which countless mono-causalities are statistically juggled in the empirical presentations of results (instead of concentrating on large correlations and discussing causalities theoretically, because no mere correlation of itself says anything meaningful about causality), and how, as it were, through the prism of preconceived classifications everything becomes more and more differentiated and evaporates into individual moments. That it would be the greater, probably also more important task to think together again what has already been separated in the tradition of conceptual preformation is hardly recognized. Complexity of the world has always confronted thinking man, especially since he increasingly no longer sees himself as part of the whole. Here lurks a source of constant error. The complexity generated by the scientific representation of the world is a different complexity from that which people encounter in everyday life.

The logic of action in everyday life has little in common with the logic of the world construction of science.[1] The often asserted fact that all human action presents itself as equally possible, blurring into one another, i.e. arbitrary (with the discourse of "postmodernism" this view has become respectable), must be regarded as extremely unenlightened, just like its supposed opposite, that it can always be traced back to an ultimate cause, a core of essence or a character. There, only the logic of everyday action is misinterpreted. Rigorous individualism, notorious self-reference, or its opposite, template-like perception, certainly obstruct an open approach to the world. A way of thinking that is free of such imponderables is open to the world and world-oriented; it neither listens only to itself, nor does it seek security in the traditional well-established, which is brought in from the outside and wants to appear with the claim of final truth. To participate in life means not least to resist doctrinal violence (Amann 2008b). At the same time, it would be beneficial to measure one's skepticism towards one's own ideas at least against those limits that doubt about other people's truth describes (Arno Plack).

---

[1] If one assumes a growing tendency towards an increase in the complexity of society, and this has been happening everywhere in sociology since Herbert Spencer, then this results in a veritable problem insofar as critical argumentation is always in danger of slipping. Social processes can become increasingly distant from scientific observation and interpretation (or vice versa), in which case what remains in science is boring rhetoric. If the difference between complex society and interpretations in everyday life is played down to the detriment of complexity, which happens constantly, for example, in the village tales cited, this amounts to social backward development or permanence of the wrong. This is mirrored in the attempts to reorganize societies in a doctrinaire way, as is currently happening in all right-wing populist regimes, because nothing is more genuine to these currents than the claim to have simple solutions for everything (similarly in Vobruba 2019).

References

Adler, A. 1979. Wozu leben wir? Frankfurt a/M.

Adorno, Th. W. 1998. Theorie der Halbbildung, in: Gesammelte Schriften. Hrsg. R. Tiedemann, Bd. 8, 93–121. Darmstadt (Lizenzausgabe).

Aldebert, H. Hrsg. 2006. Demenz verändert. Hintergründe erfassen – Deutungen finden – Leben gestalten. Schenefeld.

Alter und Zukunft: Wissen und Gestalten. Forschungsexpertise zu einem Bundesplan für Seniorinnen und Senioren. Wien 2010 (Wiener Institut für Sozialwissenschaftliche Dokumentation und Methodik. Forschungsbericht für das Bundesministerium für Arbeit, Soziales und Konsumentenschutz. Wissenschaftliche Gesamtleitung: Prof. Dr. Anton Amann).

Altner, G. 1987. Die große Kollision. Graz.

Altwerden: Altwerden in Niederösterreich. Altersalmanach 2008 von M. Bittner und G. Ehgartner. Wissenschaftliche Leitung: Prof. Dr. Anton Amann. St. Pölten 2009.

Amann, A. 1983. Lebenslage und Sozialarbeit. Elemente zu einer Soziologie von Hilfe und Kontrolle. Berlin.

Amann, A. 1989a. Die vielen Gesichter des Alters. Tatsachen, Fragen, Kritiken. Wien.

Amann, A. 1989b. Älterwerden in der bäuerlichen Welt. Notizen über eine österreichische Gemeinde im Umbruch. Forschungsbericht. Wien.

Amann, A. 1990. In den biographischen Brüchen der Pensionierung oder der lange Atem der Erwerbsarbeit, in: Die doppelte Sozialisation Erwachsener. Hrsg. E. H. Hoff, 177–204. München.

Amann, A. 1996. Soziologie. Ein Leitfaden zu Theorien, Geschichte und Denkweisen. Wien.

Amann, A. 2000. Sozialpolitik und Lebenslagen älterer Menschen, in: Lebenslagen im Alter. Gesellschaftliche Bedingungen und Grenzen, Hrsg. G. M. Backes und W. Clemens, 53–74. Opladen.

Amann, A. 2004. Die großen Alterslügen. Generationenkrieg, Pflegechaos, Fortschrittsbremse? Wien.

Amann, A. 2008a (2. Aufl. 2014). Sozialgerontologie: ein multiparadigmatisches Forschungsprogramm?, in: Das erzwungene Paradies des Alters? Fragen an eine Kritische Gerontologie. Hrsg. A. Amann, A. und F. Kolland, 45–62. Wiesbaden.

Amann, A. 2008b. Nach der Teilung der Welt. Logiken globaler Kämpfe. Wien.

Amann, A. 2011. Alter und Zukunft. Wissen und Gestalten. Forschungsexpertise zu einem Bundesplan für Seniorinnen und Senioren. Wien.http://www.bmask.gv.at/site/Soziales/Seniorinnen_und_Senioren/Teilhabe_aelterer_Menschen/, 5.11. 2012.

Amann, A. 2012. Konstruktionen des Alters. Soziale, politische und ökonomische Strategien, in: Alter(n) anders denken. Kulturelle und biologische Perspektiven. Hrsg. B. Röder, W. de Jong und K. W. Alt, 209–225. Wien-Köln-Weimar.

Amann, A. 2017. Vom guten Leben und seinen Feinden. Eine Gegenwartskritik. Wien.

Amann, A. und R. Loos. 2016. Ältere Arbeitskräfte und "Arbeit" im Alter, in: Bundesplan für Seniorinnen und Senioren. Evaluation. Wien (Büro für Sozialtechnologie und Evaluationsforschung, im Auftrag des BMASK).

Amann, A. und G. Majce. Hrsg. 2005. Soziologie in interdisziplinären Netzwerken. Wien-Köln-Leipzig.

Amann, A., G. Ehgartner und D. Felder. 2010a. Sozialprodukt des Alters. Über Produktivitätswahn, Alter und Lebensqualität. Wien.

Amann, A., G. Knapp und H. Spitzer. 2010b. Alternsforschung und Soziale Arbeit in Österreich, in: Altern, Gesellschaft und Soziale Arbeit. Lebenslagen und soziale Ungleichheit von alten Menschen in Österreich. Hrsg. G. Knapp und H. Spitzer, 553–566. Klagenfurt/Celovec-Laibach/Ljubljana-Wien/Dunaj.

Amann, A., Ch. Bischof und A. Salmhofer. 2016. Intergenerationelle Lebensqualität. Diversität zwischen Stadt und Land. Wien.

Amann, A. Ch. Bischof und I. Findenig (unter Mitarbeit von A. Fassl). 2018. Teilhabe im Alter: Theoretische Konzeptionen, praktische Gegebenheiten. Wien (Forschungsbericht für das Bundesministerium für Arbeit, Soziales, Gesundheit und Konsumentenschutz).

Aner, K. 2005. "Ich will, dass etwas geschieht!". Wie zivilgesellschaftliches Engagement entsteht – oder auch nicht. Berlin.

Aner, K. und D. Köster. 2016. Partizipation älterer Menschen – Kritisch gerontologische Anmerkungen, in: Teilhabe im Alter gestalten. Aktuelle Themen der Sozialen Gerontologie. Hrsg. G. Naegele, E. Olbermann und A. Kuhlmann, 465–483. Wiesbaden.

Arendt, H. 1981. Vita activa oder vom tätigen Leben. München.

Bachmann, I. 1996. Unter Mördern und Irren, in: Sämtliche Erzählungen, 159–186. München.

Bäcker, G. 2016. Altersarmut, Lebensstandardsicherung und Rentenniveau, in: Teilhabe im Alter gestalten. Aktuelle Themen der Sozialen Gerontologie. Hrsg. G. Naegele, E. Olbermann und A. Kuhlmann, 63–82. Wiesbaden.

Bähr, J. 1983. Bevölkerungsgeographie. Stuttgart.

Bayertz, K. 2012. Der aufrechte Gang. Eine Geschichte des anthropologischen Denkens. München.

Beauvoir, S. 1977. Das Alter. Reinbek/Hamburg.

Berelson, B. und G. A. Steiner. 1972. Menschliches Verhalten. Bd. II: Soziale Aspekte. Weinheim.

Bergener, M. und S. I. Finkel, Hrsg. 1995. Treating Alzheimer's and Other Dementias. New York.

Blumenberg, H. 1979. Arbeit am Mythos. Frankfurt a/M.

Blumenberg, H. 2010. Theorie der Lebenswelt. Frankfurt a/M.

Bongaarts, J. 2005. Long-Range Trends in Adult Mortality. Models and Projection Methods, in: Demography 42/1, 23–49.

Born, K. E., H. Henning und F. Tennstedt. Hrsg. 2000a. Quellensammlung zur Geschichte der Deutschen Sozialpolitik 1867 bis 1881. I. Abteilung, 7. Band (Erster Halbband): Armengesetzgebung und Freizügigkeit. Mainz.

Born, K. E., H. Henning und F. Tennstedt. Hrsg. 2000b. Quellensammlung zur Geschichte der Deutschen Sozialpolitik 1867 bis 1914. I. Abteilung, 7. Band (Zweiter Halbband): Armengesetzgebung und Freizügigkeit. Mainz.

Borscheid, P. 1989. Geschichte des Alters. München.

Brock, B. 1998. Warum diese Ausstellung, in: Die Macht des Alters. Strategien der Meisterschaft. Hrsg. B. Brock. Katalog zur gleichnamigen Ausstellung. Köln.

Brunner, O. 1956. Das "ganze Haus" und die alteuropäische Ökonomik, in: Neue Wege der Verfassungs- und Sozialgeschichte. Göttingen.

Bundesministerium für Familie, Senioren, Frauen und Jugend (BMFSFJ). 2019. Frauen und Männer in der zweiten Lebenshälfte – Älterwerden im sozialen Wandel. Zentrale Befunde des Deutschen Alterssurveys (DEAS) 1996 bis 2017. Berlin.

Bundesplan für Seniorinnen und Senioren: Evaluation. 2016. Endbericht. Büro für Sozialtechnologie und Evaluationsforschung (Wiss. Leitung Prof. Dr. Anton Amann). Wien.

Burgess, E. W. 1962. Western European Experience in Aging as Viewed by an American, in: Social Welfare of the Aging. Hrsg. J. Kaplan und J. Aldridge. New York-London.

Butor, M. 1984. Der Roman als Sache, in: Alchemie und ihre Sprache. Hrsg. M. Butor, 53–60. Frankfurt a/M.

Campbell, D. und T. Barth. 2009. Wie können Demokratie und Demokratiequalität gemessen werden? Modelle, Demokratie-Indices und Länderbeispiele im globalen Vergleich, in: Sozialwissenschaftliche Rundschau 49, 209–231.

Camus, A. 1950. Die Pest. Reinbek/Hamburg.

Caplan, R. D. 1987. Person-environment fit theory and organizations: Commensurate dimensions, time perspectives, and mechanisms, in: Journal of Vocational Behavior 31, 248–267.

Clemens, W. 2010. Lebensläufe im Wandel – Gesellschaftliche und sozialpolitische Perspektiven, in: Soziale Lebenslaufpolitik. Hrsg. G. Naegele, 87–109. Wiesbaden.

Dahrendorf, R. 1975. Die neue Freiheit. Überleben und Gerechtigkeit in einer veränderten Welt. München.

Diamond, J. 2013. Vermächtnis. Was wir von traditionellen Gesellschaften lernen können. Frankfurt a/M.

Douglas, A. 1971. The Feminization of American Culture. New York.

Dux, G. 1992. Die Spur der Macht im Verhältnis der Geschlechter. Über den Ursprung der Ungleichheit zwischen Mann und Frau. Frankfurt a/M.

Dux, G. 2013. Demokratie als Lebensform. Die Welt nach der Krise des Kapitalismus. Weilerswist.

Dux, G. 2017. Die Evolution der humanen Lebensform als geistige Lebensform. Handeln-Denken-Sprechen. Wiesbaden.

Ebner-Eschenbach, M. v. 1985. Das Gemeindekind, in: Ausgewählte Erzählungen. Hrsg. P. Friedländer, Bd. 2, 5–201. Berlin.

Eiffe, F. F., M. Till, G. Datler et al. 2012. Soziale Lage älterer Menschen in Österreich. Sozialpolitische Schriftenreihe des BMASK, Bd. 11. Wien.

Eilenberger, W. 2018⁹. Zeit der Zauberer. Das große Jahrzehnt der Philosophie 1919–1929. Stuttgart.

Elwert, G. 1992. Alter im interkulturellen Vergleich, in: Zukunft des Alterns und gesellschaftliche Entwicklung. Hrsg. Baltes, P. B. und J. Mittelstraß, 260–282. Berlin-New York.

Erikson, E. H., J. M. Erikson und H. Q. Kivnick 1986. Vital Involvement in Old Age. New York-London.

Erlinghagen, M. 2008. Ehrenamtliche Arbeit und informelle Hilfe nach dem Renteneintritt. in: Produktives Altern und informelle Arbeit in modernen Gesellschaften. Theoretische Perspektiven und empirische Befunde. Hrsg. Erlinghagen, M. und K. Hank. Wiesbaden.

Fagan, B. 2012. Cro-Magnon. Das Ende der Eiszeit und die ersten Menschen. Stuttgart.

Findenig, I. 2017. Generationenprojekte. Orte des intergenerativen Engagements. Potenziale, Probleme und Grenzen. Opladen-Berlin-Toronto.

Franz, J. 2009. Intergenerationelles Lernen ermöglichen: Orientierungen zum Lernen der Generationen in der Erwachsenenbildung. Bielefeld.

Friedländer, P. 1985. Nachwort, in: Ebner-Eschenbach, M. v., Ausgewählte Erzählungen. Hrsg. P. Friedländer, Bd. 1, 5–17. Berlin.

Fries, J. F. 1983. The Compression of Morbidity, in: Milbank Quarterly 61/3, 397–419.

Gadamer, H.-G. und P. Vogler. Hrsg. 1975. Philosophische Anthropologie. Bde. 6 und 7 der Reihe "Neue Anthropologie". Stuttgart.

Gehlen, A. 1986a. Der Mensch. Wiesbaden.

Gehlen, A. 1986b. Urmensch und Spätkultur. Wiesbaden.

Gerhardt, V. 1999. Selbstbestimmung. Das Prinzip der Individualität. Stuttgart.

Geulen, D. 1989. Das vergesellschaftete Subjekt. Frankfurt a/M.

Giddens, A. 1995. Konsequenzen der Moderne. Frankfurt a/M.

Gilleard, Ch. und P. Higgs. 2000. Cultures of Ageing. London.

Glascock, A. O. 1986. Resource Control Among Older Males in Southern Somalia, in: Journal of Cross-Cultural Gerontology 1/1, 51–72.

Goethe, J. W., v. 1981. Maximen und Reflexionen: Nr. 1099, in: Werke, Hamburger Ausgabe. Komm. v. E Trunz, Bd. 12. München.

Gollner, H. 2012. Adalbert Stifter, in: Eine Literaturgeschichte: Österreich seit 1650. Hrsg. K. Zeyringer und H. Gollner, 327–351. Innsbruck-Wien-Bozen.

Gronau, M. 2016. Politische Aporetik als Ausweg aus der Ausweglosigkeit. Paradigmen der griechischen Antike, in: Politische Aporien. Akteure und Praktiken des Dilemmas. Hrsg. M.-L. Frick und Ph. Hubmann, 23–50. Wien-Berlin.

Habermas, J. 1970. Anthropologie, in: Arbeit, Erkenntnis, Fortschritt. Hrsg. J. Habermas, Amsterdam, 164–180.

Hafner-Fink, M., S. Kurdija und S. Uhan. 2017. Social Research: From Paradigmatic Divide to Pragmatic Eclecticism. Wien.

Hampel, H. und J. Pante. 2008. Aktuelle Frühdiagnostik der Alzheimer Demenz, in: Neurotransmitter 19, 26–32.

Harari, Y. N. 2015. Eine kurze Geschichte der Menschheit. München.

Hartenstein, W. und G.-U. Weich. 1993. Mobilität und Verkehrsmittelwahl, in: Bericht und Dokumentation zum Fachkongress "Verkehrssicherheit älterer Menschen." Hrsg. B. Schlag. Bonn.

Hauser, A. 1973. Kunst und Gesellschaft. München.

Hauser, A. 1988. Soziologie der Kunst. München.

Heimgartner, A. und I. Findenig. 2017. Biografie und freiwilliges Engagement. Wien (Forschungsbericht für das Bundesministerium für Arbeit, Soziales und Konsumentenschutz).

Heinze, R. G., G. Naegele, und K. Schneiders. 2011. Wirtschaftliche Potentiale des Alters. Stuttgart.

Henning, H. und F. Tennstedt. Hrsg. 2003. Quellensammlung zur Geschichte der Deutschen Sozialpolitik 1867 bis 1914. II. Abteilung, 1. Band: Grundfragen der Sozialpolitik. Hrsg. H. Henning und F. Tennstedt. Mainz.

Herder, J. G. O. J. Ideen zur Philosophie der Geschichte der Menschheit. Herders Werke. Hrsg. H. Kurz, Bd. 3 (nach der Ausgabe von 1784). Leipzig-Wien.

Hobsbawm, E. 2004. Das Gesicht des 21. Jahrhunderts. München.

Höfler, S., Th. Bengough, P. Winkler und R. Griebler. Hrsg. 2015. Österreichischer Demenzbericht. Bundesministerium für Gesundheit und Sozialministerium. Wien. (www.bmfg. gv.at).

Hüther, M. und G. Naegele. Hrsg. 2013. Demografiepolitik. Herausforderungen und Handlungsfelder. Wiesbaden.

Jenny, M. 1996. Psychische Veränderungen im Alter. Mythos-Realität-Psychologische Interventionen. Wien.

Jonas, H. 1980. Das Prinzip Verantwortung. FRankfurt a/M.

Kelle, U. 2008. Alter & Altern, in: Handbuch Soziologie. Hrsg. N. Baur, H. Korte, M. Löw und M. Schroer, 11–31. Wiesbaden.

Kidahashi, M. und R. J. Manheimer. 2009. Getting Ready for the Working-in-Retirement Generation. How Should LLIs Respond?, in: The LLI Review 4, 1–8.

Kment, A. 1996. On the Ontology of Biological Aging, in: Vitality, Mortality and Aging. Hrsg. A. Viidik und G. Hofecker, 5–11. Wien.

Kohli, M. 1978. Soziologie des Lebenslaufs. Darmstadt-Neuwied.

Kohli, M. 1985. Die Institutionalisierung des Lebenslaufs. Historische Befunde und theoretische Argumente, in: Kölner Zeitschrift für Soziologie und Sozialpsychologie 37/1, 1–29.

Kolland, F. 1996. Kulturstile im Alter. Wien.

Kolland, F. 2016. Bildungsmotivation im Alter. Modelle und Forschungserkenntnisse. Forschungsbericht. Hrsg. Bundesministerium für Arbeit, Soziales und Konsumentenschutz. Wien.

Kolland, F. und A. Amann. 2013. Alter und Altern, in: Forschungs- und Anwendungsfelder der Soziologie. Hrsg. E. Flicker und R. Forster, 30–45. Wien.

Kraus, K. 1914–1918. Die Fackel, Bd. 7, Nr. 398 bis 507. April 1914 bis Januar 1919. München.

Kraus, K. 1968–1976. Die Fackel, Bd. 7, Nr. 398 bis 507, April 1914 bis Januar 1919. München (Neuausgabe Frankfurt a/M o. J.).

Kricheldorff, C. 2010. Bildungsarbeit mit älteren und alten Menschen, in: Handbuch Soziale Arbeit und Alter. Hrsg. K. Aner und U. Karl, 99–109. Wiesbaden.

Krüger, H. 2010. Familienpolitik und Lebenslaufforschung miteinander verknüpfen: ein zweifacher Gewinn, in: Soziale Lebenslaufpolitik. Hrsg. G. Naegele, 217–244. Wiesbaden.

Kuhlmann, A. G. Naegele und E. Olbermann 2016. Einführung, in: Teilhabe im Alter gestalten. Aktuelle Themen der Sozialen Gerontologie. Hrsg. G. Naegele, E. Olbermann und A. Kuhlmann, 45–60. Wiesbaden.

Kuzmics, H. und G. Mozetič. 2003. Literatur als Soziologie. Zum Verhältnis von literarischer und gesellschaftlicher Wirklichkeit. Konstanz.

Laslett, P. 1995. Das dritte Alter. Historische Soziologie des Alterns. München.

Lawton, M. P. 1990. Residential environment and self-directedness among older people, in: American Psychologist 45/5, 638–640.

Lee, R. und R. Daly. Hrsg. 1999. The Cambridge Encyclopedia of Hunters and Gatherers. Cambridge.

Lehr, U. 1991. Psychologie des Alterns. Heidelberg.

Leibfried, St. L. Leisering, P. Buhr et al. 1995. Zeit der Armut. Lebensläufe im Sozialstaat. Frankfurt a/M.

Lepenies, W. 1971. Anthropologie und Gesellschaftskritik, in: Kritik der Anthropologie. Hrsg. W. Lepenies und H. Nolte, 77–102. München.

MacGregor, N. 2017. Eine Geschichte der Welt in 100 Objekten. München.

Magris, C. 1966. Der habsburgische Mythos in der österreichischen Literatur. Salzburg (ital. 1963).

Malnar, B. und K. H. Müller 2015. Surveys and Reflexivity. A Second-Order Analysis of the European Social Survey (ESS). Wien.

Manton, K. G., E. Stallard und L. S. Corder. 1998. The Dynamics of Dimensions of Age-related Disability 1982 to 1994 in the U.S. Elderly Population, in: The Journal of Gerontology. Series A, Biological Sciences and Medical Sciences 53/1, 59–70.

Maturana, H. R. und F. Varela 1987. Der Baum der Erkenntnis. Bern-München-Wien (erstmals 1984).

Mayer, K.-U. 1990. Lebensverläufe und gesellschaftlicher Wandel. Opladen.

Menning, S. 2009. Wahlverhalten und politische Partizipation älterer Menschen. Hrsg. Deutsches Zentrum für Altersfragen Berlin (Report Altersdaten 3/2009). URN:http://nbn-resolving.de/urn:nbn:de:0168-ssoar-370243.

Minois, G. 1987. Histoire de la vieillesse en Occident de L'Antiquité à la Renaissance. Paris.

Naegele, G. 2010. Soziale Lebenslaufpolitik – Grundlagen, Analysen und Konzepte, in: Soziale Lebenslaufpolitik. Hrsg. G. Naegele, 27–85. Wiesbaden.

Nahnsen, I. 1975. Bemerkungen zum Begriff und zur Geschichte des Arbeitsschutzes, in: Arbeitssituation, Lebenslage und Konfliktpotential. Festschrift für Max E. Graf zu Solms-Roedelheim. Hrsg. M. Osterland, 145–166. Frankfurt-Köln.

Neugarten, B. L. 1974. Age Groups in American Society and the Rise of the Young Old, in: Annals of the Academy of Social and Political Science 41/5, 187–198.

Neurath, O. 1981. Grundlagen der Sozialwissenschaften, in: Otto Neurath: Gesammelte philosophische und methodologische Schriften. Hrsg. R. Haller und H. Rutte, Bd. 2, 925–978. Wien (zuerst 1944).

Nietzsche, F. 1980. Vom Nutzen und Nachteil der Historie für das Leben, in: Werke, Bd. 1. Hrsg. K. Schlechta. München.

Oswald, F. und H.-W. Wahl. 2016. Alte und neue Umwelten des Alters – Zur Bedeutung von Wohnen und Technologie für die Teilhabe in der späten Lebensphase, in: Teilhabe im Alter gestalten. Aktuelle Themen der Sozialen Gerontologie. Hrsg. G. Naegele, E. Olbermann und A. Kuhlmann, 113–129. Wiesbaden.

Pantel, J. 2017. Alzheimer-Demenz von Auguste Deter bis heute. Fortschritte, Enttäuschungen und offene Fragen, in: Zeitschrift für Gerontologie und Geriatrie 50/7, 576–587.

Pascal, B. O. J. Gedanken. Nach der endgültigen Ausgabe übertragen von W. Rüttenauer. Biersfelden-Basel (Sammlung Dieterich).

Plack, A. 1979. Philosophie des Alltags. Stuttgart.

Plessner, H. 1975³. Die Stufen des Organischen und der Mensch. Berlin-New York.

Plessner, H. 1985. Schriften zur Soziologie und Sozialphilosophie, Gesammelte Schriften Bd. X. Hrsg. G. Dux, O. Marquard und E. Ströker. Frankfurt a/M.

Popper, K. 1970. Der Zauber Platons. Die offene Gesellschaft und ihre Feinde, Bd. 1. Bern-München.

Portmann, A. 1969. Biologische Fragmente zu einer Lehre vom Menschen. Basel.

Radebold, H., H. Bechtler und I. Pina.1981. Therapeutische Arbeit mit älteren Menschen. Freiburg i/B.

Radebold, H. und H. Radebold. 2009. Älterwerden will gelernt sein. Stuttgart.

Rehberg, K.-S. 1986¹³. Arnold Gehlens Beitrag zur "Philosophischen Anthropologie". Einleitung in die Studienausgabe seiner Hauptwerke, in: A. Gehlen: Der Mensch, I-XVII. Wiesbaden.

Riley, M. W. und J. Riley. 2000. Age Integration. Conceptual and Historical Background, in: The Gerontologist 40/3, 266–270.

Rosenmayr, L. 1974. Die Revision der These vom generellen Leistungsverfall im Alternsprozeß, in: Aktivitätsprobleme des Alternden. Eine psychosomatische Studie. Hrsg. K. Fellinger, 101–123. Basel.

Rosenmayr, L. 1990. Die Kräfte des Alters. Wien.

Ruppe, G. und A. Stückler. 2015. Österreichische Interdisziplinäre Hochaltrigenstudie. Zusammenwirken von Gesundheit, Lebensgestaltung und Betreuung. Wien (Forschungsbericht).

Ryan, Ch. und C. Jethá. 2010. Sex at Dawn. The Prehistoric Origins of Modern Sexuality. New York.

Ryan, L. H., J. Smith, T. C. Antonucci und J. S. Jackson. 2012. Cohort Differences in the Availability of Informal Caregivers. Are the Boomers at Risk? in: The Gerontologist 52/2, 177–188.

Schaub, R. und H. v. Lützau-Hohlbein. 2017. Demenz – Sicht der Betroffenen und Angehörigen, in: Zeitschrift für Gerontologie und Geriatrie 50/7, 616–622.

Scheler, M. 1954⁴. Der Formalismus in der Ethik und die materiale Wertethik. Bern.

Scheler, M. 1988¹¹. Die Stellung des Menschen im Kosmos. Bonn 1988 (erstmals 1928).

Schimank, U. 2000². Theorien gesellschaftlicher Differenzierung. Opladen.

Schütze, Y. 1982². Psychoanalytische Theorien in der Sozialisationsforschung, in: Handbuch der Sozialisationsforschung. Hrsg. K. Hurrelmann und D. Ulich, 15–49. Weinheim-Basel.

Schulz-Nieswandt, F. 2007. Lebenslauforientierte Sozialpolitikforschung, Gerontologie und philosophische Anthropologie. Schnittflächen und mögliche Theorieklammern, in: Altersforschung am Beginn des 21. Jahrhunderts. Hrsg. H.-W. Wahl und H. Mollenkopf, 61–81. Berlin.

Schwenk, O. 1999. Soziale Lagen in der Bundesrepublik. Opladen.

Segalen, V. 1983. Die Ästhetik des Diversen. Aufzeichnungen. Frankfurt a/M-Paris.

Sieder, R 1987. Sozialgeschichte der Familie. Frankfurt a/M.

Simmons, L. 1945. The Role of the Aged in Primitive Societies. New Haven.

SIZE. 2006. Life Quality of Senior Citizens in Relation to Mobility Conditions. Final Report (Deliverable D18) of a Project in the EU's Fifth Framework Programme. University of Vienna (by Anton Amann and Barbara Reiterer).

Soeffner, H.-G. 1989. Auslegung des Alltags – Alltag der Auslegung. Frankfurt a/M.

Sombart, W. 1902. Der moderne Kapitalismus, Bd. II. Leipzig.

Stifter, A. 1951. Werke in zwei Bänden. Hrsg. I. E. Walter. Salzburg-Stuttgart.

Sting, S. 2010. Soziale Bildung, in: Enzyklopädie Erziehungswissenschaft Online (EEO), Fachgebiet Soziale Arbeit. Hrsg. W. v. Schröer, und C. Schweppe. Weinheim-München.

Streminger, G. 1994. David Hume. Sein Leben und sein Werk. Paderborn-München-Wien-Zürich.

Strigl, D. 2016. Berühmt sein ist nichts. Marie von Ebner-Eschenbach – Eine Biographie. Wien-Salzburg.

Strigl, D., E. Polt-Heinzl und U. Tanzer. 2015. Marie von Ebner-Eschenbach – Leseausgabe. Wien-Salzburg.

Sutter, T. 2003. Sozialisationstheorie und Gesellschaftsanalyse. Zur Wiederbelebung eines zentralen soziologischen Forschungsfeldes, in: Subjekte und Gesellschaft. Zur Konstitution von Sozialität. Hrsg. U. Wenzel, B. Bretzinger und K. Holz, 45–69. Weilerswist.

Tews, H.-P. 1993. Neue und alte Aspekte des Strukturwandels des Alters, in: Lebenslagen im Strukturwandel des Alters. Hrsg. G. Naegele und H.-P. Tews, 15–42. Opladen.

Tocqueville, A. de. 2016. Über die Demokratie in Amerika. Stuttgart (erstmals 1835).

Tomasello, M. 2013. Eine Naturgeschichte menschlichen Denkens. Frankfurt a/M.

Topp, H. H. 2001. Kürzere Wege, mehr Mobilität, weniger Verkehr, in: Ökologische Stadtentwicklung: Innovative Konzepte für Städtebau, Verkehr und Infrastruktur. Hrsg. M. Koch. Stuttgart.

UNECE. 2010. PolicyBrief: Integration und Teilhabe älterer Menschen in der Gesellschaft. UNECE Kurzdossier zum Thema Altern Nr. 4, Dezember (ECE/WG.1/4).

Vobruba, G. 2019. Aufklärung: der Theorie und der Leute, in: Aufklärung als Aufgabe der Geistes- und Sozialwissenschaften. Beiträge für Günter Dux. Hrsg. U. Bröckling, und A. Paul, 106–111. Weinheim-Basel.

Weber, M. 1922. Wirtschaft und Gesellschaft. Tübingen.

White, H. 1981. The Value of Narrativity in the Representation of Reality, in: On Narrative. Hrsg. W. J. Th. Mitchell, 1–23. Chicago.

WHO. 2002. Active ageing. A Policy Framework. Geneva, Switzerland: World Health Organization.

Wolfgruber, G. 1985. Die Nähe der Sonne. Wien-Salzburg.

Zeyringer, K. und H. Gollner. 2012. Eine Literaturgeschichte: Österreich seit 1650. Innsbruck-Wien-Bozen.

Zinnecker, J. 2000. Selbstsozialisation – Essay über ein aktuelles Konzept, in: Zeitschrift für Sozialisationsforschung und Erziehungssoziologie 20/3, 272–290.

The manufacturer's authorised representative in the EU is Springer
Nature Customer Service Centre GmbH, Europaplatz 3, 69115 Heidelberg,
Germany. If you have any concerns regarding our products, please
contact ProductSafety@springernature.com

Printed and bound by CPI Group (UK) Ltd, Croydon, CR0 4YY

28/04/2026

02098512-0002